opening page: **Astronaut firing his maneuvering unit, with space station and shuttlecraft below.**

preceding page: **Lift-off on Titan, the largest satellite of the planet Saturn, seen looming above. The graying of the sky above the horizon indicates Titan's atmosphere, tenuously held, but the only atmosphere detected on any of the satellites in our solar system.**

this page, above: **Open cargo hatch of the first Mars lander, detail of painting on pages 116-17.**

below: **Control tower on Europa, one of the twelve moons of Jupiter and about the same size as Earth's moon; detail of painting on pages 138-39.**

Our World
in Space

Our World in Space

Robert McCall · Isaac Asimov

Foreword by Edwin E. Aldrin, Jr.

New York Graphic Society Ltd.
Greenwich, Connecticut

Frontispiece: Exploring the moon in advanced
lunar rovers in the late 1980's.

International Standard Book Number 0-8212-0434-3
Library of Congress Catalogue Number 73-78567

First published 1974 by New York Graphic Society, Ltd.,
140 Greenwich Avenue, Greenwich, Conn. 06830
First printing 1974

Designed by Betsy Beach
Color photography by Frank Lerner
Printed and bound in Italy by Mondadori Editore, Verona

Contents

Lunar survey system. The astronaut uses a hypothetical advanced surveyor's instrument.

Foreword

by Edwin E. Aldrin, Jr.

Traditionally, man is impatient of the immediate future. He tends to want, to hope, above all, to imagine, more than is humanly or technologically possible in his flicker of time. But the events of the past decade have confirmed the soundness of that yearning instinct. They have lifted the hearts of the dreamers, and of all but the most Earth-bound pessimists.

Musing over the well-founded prognostications in the remarkable paintings and text of this volume, I found myself considering—as I have so often in the past four years—the meaning of the unique events I took part in and my own role in them. I was asked to participate in this publication, I suppose, because I was the winner of an extraordinary lottery: I was selected to take part in the first Moon landing, an experience that became symbolic for all Earth men.

The spectacular voyage of Apollo 11 meant the conquering of a new frontier. Two men, carrying with them the collective dreams of mankind throughout the ages, set foot on the Moon. It was a triumph of imagination and of collective technology. In a troubled, disheartened time, the voyage was a glorious witness to man's ability to create, to realize his dreams.

Twelve men have now walked on the Moon. The twelve of us have, I think, at least one common viewpoint: We share a special concept of the Earth as a planet. We have looked on it from the surface of the Moon and seen it whole in space—beautiful, bright, not very large . . . and somehow vulnerable. To me, it seemed a great place to come from, and an even greater place to return to. Millions now have had a reflection of this view of Earth through the dramatic photographs taken in space; I think they too begin to share this sense of planet Earth as a place to cherish, and even are beginning to act upon it.

The collective technology behind the triumph of Apollo 11 can be translated in terms of the absorption of the skills of thousands of highly trained and talented people over a period of ten years, and the expenditure of some twenty-four billion dollars. All this, to achieve a dream. The space program has long had its critics, and since 1969 (perhaps, ironically, because of the apparently routine success of its missions) it has lost some of its hold on the public imagination. The verification of man's inventiveness demonstrated in the first Skylab mission and the economy-oriented evolution of the space shuttle may convince some critics of NASA and the space industry, but others are still unsatisfied. Even as we casually or unknowingly profit from some of the fringe benefits already accrued from the program—such as extreme electronic miniaturization, satellite communications, new surgical techniques, improved weather forecasting—there are those who consider the entire program a wasteful, even useless, expenditure of talent and money. I can never agree. The voyage to the Moon will have been useless only if we do not use the experience. And I believe we have already begun to do so both in technical and in spiritual terms.

If you think of the evolution of the space program, you may be as startled as I occasionally am by the rush of technology and history. After Yuri Gagarin's one-orbit flight outside the Earth's atmosphere in 1961, the space program grew by such quantum leaps that the mind blurs: the first American program, Mercury, quickly evolved into Gemini, twelve flights in preparation for an eventual lunar landing. Then came Apollo, and on July 20, 1969, only eight years after the first manned space flight, men walked on the Moon. Our training and knowledge were so extensive that there were virtually no unknowns or unanticipated events during the entire voyage of Apollo 11. We went all the way from Earth to the Moon and back to Earth within half a second of the time allotted by our flight plan. Even the Moon was much as we expected to find it.

Man has increased his knowledge of the unknown so rapidly and effectively in recent years that he can now envision realistically what it is like where he has yet to go. The images in this volume are not fanciful dreams, but have a basis in reality; they are expressions of some of the imaginative strivings toward the future. The ingenuity and imagination of the individual—or collective—mind of man is virtually limitless. Yet in long-range terms the technological resources available to him may make it possible to achieve, even surpass, his wildest flights of fancy. And as man develops the tools and capabilities to extend his reach farther and farther, there is no doubt he will feel compelled to go as far as he is capable of going.

Buzz Aldrin

page 13, above: Dream of another reality, detail of painting on pages 174-75.

page 13, below: Astronaut entering recovery port of a space station, detail of painting on pages 144-45.

preceding page: Space station in Earth orbit, perhaps a hundred years hence. The transparent hemisphere shelters a nuclear power facility, an observation and control center is at right, and inhabited maneuverable spheres are in left background.

this page: A schematic view of our solar system, with the nine planets depicted in their relative sizes and the Sun at left showing Sun spots and some of the surface prominences. In the center is our Milky Way galaxy, with the red arrow pointing to our solar system within it.

1 The Moon

The most surprising thing about the Moon is that it is there at all.

Anyone studying the planets of the Solar system would guess that the Earth has no satellite at all—or at most one or two tiny ones.

It is not surprising that large planets with large gravitational fields should have many satellites held in the grip of those fields. The four largest planets of the Solar system, Jupiter, Saturn, Uranus, and Neptune, have twenty-nine satellites altogether (possibly more, but twenty-nine is all we have discovered so far). Of these twenty-nine, only six have diameters in the range of two to three thousand miles and may be thought of as giant satellites. The rest are all smaller; some are only ten or fifteen miles across.

Of the giant satellites, four circle Jupiter, the largest of all the planets. They are Io, Europa, Ganymede, and Callisto. A fifth, Titan, circles Saturn, while a sixth, Triton, whirls about Neptune. Large though these satellites are, they are small compared to the giant planets about which they revolve. Ganymede, the largest of the satellites, has only 1/12,000 the mass of Jupiter. Titan has only 1/4,100 the mass of Saturn. Triton has only 1/760 the mass of Neptune. The largest of Uranus's five satellites is Titania, which with a diameter of about 1,000 miles is no giant and has only about 1/25,000 the mass of the planet it circles.

We might judge from this that even the largest satellite is rarely more than 1/1000 the mass of the planet it circles and is much more likely to be 1/5,000 the mass or less. If this is so, we might expect the Earth to have a satellite with a diameter of no more than 500 miles, *at most*.

Next, suppose we consider the four small planets of the Solar system (disregarding the Earth), Pluto, Mars, Venus, and Mercury. Of these, Pluto has no known satellite, but it is tremendously far away from us and not much is known about it, so perhaps we should not give too much weight to its evidence.

Mercury and Venus, however, are not very far from us and we can see them clearly. They have no satellites; that is quite certain. Nothing of any respectable size, nothing even a couple of miles across, circles either planet. And Venus is very nearly the size of Earth.

Mars, which is smaller than either Venus or Earth but larger than Mercury, does have satellites—two, in fact. The Martian satellites, however, are tiny things. One is a little over a dozen miles across, the other a little less.

Based on our knowledge of the other small planets, we might decide that even a 500-mile-wide satellite is far too large to expect for Earth. We would surely feel safe in deciding that Earth would have no satellite at all, or at best one so small that it would appear as no more than a moving star when seen from Earth's surface.

Yet this is not so.!

Earth *has* a satellite; and as it happens, it is a giant. The Moon is 2,160 miles in diameter and would be a respectable object even if it were circling Jupiter. It has 1/81 the mass of Earth. It is far larger in comparison to the planet it circles—*far* larger—than any other satellite in the Solar system.

Astronomers have not been able to explain why the earth should have such an outsize satellite, and some suspect it may not have been a satellite at all to begin with. It may have been an independent planet which, passing close to Earth occasionally, was eventually captured, becoming a satellite.

Be that as it may, does it matter? Is there any *real* importance to the fact that the Earth has a giant satellite, aside from the fact that it looks pretty and sometimes casts a dim light at night?

Perhaps. On at least one occasion during the history of life, the existence of the Moon may have been of supreme importance. It may, in fact, have led to the existence of man.

Life originated in the sea, and until about 400 million years ago remained confined to the sea. The land surface of the planet remained desolate. For plant and animal life to conquer the land, a long and slow process of adaptation and acclimation must have taken place, and this might not have happened (or at least not as easily or as well) were it not for the Moon.

The Moon was in the sky at the beginning of the conquest of the land. Even if the Moon was captured and has not always circled the Earth, its capture could not have been very recent and probably took place at least a billion years ago.

The Moon in the sky means tides in the ocean. The gravitational effect of a large body like the Moon so close to the Earth (and it seems certain that the Moon was closer to the Earth several hundreds of millions of years ago than it is now and therefore would have had a stronger gravitational effect then than now) would produce a bulge of water on either side of the planet. Each bulge would pass over

18

a given shoreline once a day as the Earth rotated, so that, in general, there would be two high tides and two low tides a day.

Twice a day the water would creep up the shore, each time carrying life-forms with it. Each time the water would fall back, taking most of the life-forms, but some would remain behind in the damp sand. Those life-forms that could survive such an ordeal would now be safer from their enemies and would flourish. Thus, plants came to grow on land and to develop woody stalks to support themselves against the pull of gravity, and roots to search for the now-distant water.

And, of course, once plants invaded the land, animals would follow.

If the Earth had had no Moon, there would still have been the Sun, which also produces tides. These might have provided the push required for the colonization of the land—but it would have been a much weaker push, for the Sun's tides are one-third the size of the Moon's, and unlike the Moon's, were no greater in the past than they are now.

A second significant service performed by the Moon was to stimulate the intellectual development of mankind. When man's brain developed to the point where he was capable of abstract thought, he may have begun to conceive of time. Yet the concept of time in itself would have been nothing but a vague thought if some technique of measuring it had not evolved.

To measure time, some regular periodic motion must be used. There are four obvious periodic motions in nature that proceed without apparent change generation after generation. These are the semi-daily rise and fall of the tides; the daily alternation of day and night; the monthly cycle of the Moon's phases; and the yearly cycle of the seasons. Of these, the cycle of the Moon's phases is particularly interesting. Its cycle is longer than those of the tide or the day, which are perhaps too short for practical purposes, but it is shorter than that of the year, which is too long. This practical attraction of the Moon is attended by the beauty and fascination of a luminous body that appears in various parts of the sky, in varying shapes, and is occasionally eclipsed.

Even very primitive men tried to count off the days from new Moon to new Moon; to fit them into the cycle of the seasons; to work out methods of predicting the next eclipse. Alexander Marshak has recently presented evidence to show that markings on ancient bones are primitive calendars; Gerald Hawkins has found in the circle at Stonehenge a kind of primitive observatory for the prediction of lunar eclipses. The Moon and its phases may have driven mankind 19 to the invention of calendars, of mathematics and astronomy, and, therefore, of science.

Finally, the Moon has served a third purpose in historic times, simply by being there, and by being as large as it is.

Early man, studying the heavens, had to be aware of the Sun first. It could scarcely be ignored. It was an unbearably bright circle of light one could not look at directly without extreme discomfort. It was a clear and obvious source of light and heat, and, without question, was utterly different in nature from the Earth.

At night there were specks of light, dots rather than circles, and easily examined without pain or damage to the eyes. Some of them, indeed, were so dim as to be barely visible. Yet all these stars (and planets, as some of the brighter ones were called) gleamed with light. The Sun and all the stars and planets by night were luminous bodies and were, in this respect, apparently fundamentally different from the Earth, which was a dark body except where it was illuminated by the Sun.

If this were all there were to the skies, men would be justified in assuming the Earth to be a unique world, alone fit for the habitation of man; all the objects in the sky served only for its illumination, and, therefore, were of no importance but as lamps to all-important man.

But there was one exception to all this—the Moon.

The Moon was a large body in appearance, as large as the Sun. Yet it was far dimmer than the Sun and could be looked at, not only without pain, but with much pleasure. Furthermore, it changed shape from night to night, and by Babylonian times at the latest, sky-watchers (''astronomers'') were sophisticated enough to see that the change in shape was not an actual one. The Moon was always a circle, as it clearly seemed to be at the time of the full Moon, but not all the circle gleamed. At times only a thin crescent shone with light, and then the rest of the circle could sometimes be made out in a very faint red glimmer (''the old Moon in the new Moon's arms'').

From the position of the lighted portion of the Moon's surface with respect to the Sun, it was obvious that the Moon was shining by the Sun's reflected light; and that the change of the phases could be associated with the changing position of the Moon with reference to the Sun, as both appeared to circle the Earth.

But that meant that the Moon itself was a dark body, like the Earth. The Earth was *not* unique, and the Moon was, in this respect at least, another Earth.

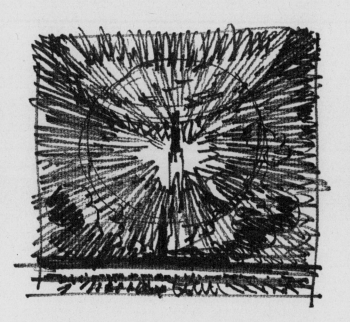

Something else easily noted by the unaided eye was the fact that the Moon alone, of all the heavenly bodies, was not featureless. The stars were all points of light of varying degrees of brightness, undistinguishable beyond that. The Sun was a circle of light, and of uniform and featureless brightness. The Moon, on the other hand, bore markings that were particularly prominent at the time of the full Moon. This attracted attention and gave rise to legends, of which the best known in our own time concerns "the man in the Moon." Even the legends gave a flavor of world-ness to the Moon: There could be a man in it.

If, then, the Moon were absent from the heavens, mankind might never, perhaps, have felt the urge to go adventuring into space—for there would have seemed to be nothing there but lights. The Moon, by its presence in the sky, gave man's unaided eye clear evidence that it was another world, presumably inhabitable. Given the restless mind and imagination of man, what else must follow but to dream of reaching the Moon?

But if the Moon were a world, how large a world could it be? To the eye it looks small, and typical ancient legend represented both Sun and Moon as chariots or ships manipulated by gods. They were no larger than human artifacts, in other words, and could not be compared in size to mountains, let alone to the Earth.

Indeed, the Greeks discovered that the Earth was even more enormous than casual inspection might lead people to think it was. About 240 B.C., the Greek philosopher Eratosthenes used trigonometric methods for determining (correctly) that the Earth was 8,000 miles in diameter.

About 130 B.C., another Greek, the astronomer Hipparchus of Nicaea, the greatest of the ancient astronomers, used trigonometric methods for determining the distance of the Moon from Earth and calculated it (correctly) to be equal to 30 times the diameter of the Earth.

This means that the Moon is about 240,000 miles from the Earth. It also means that the Moon, to appear as large as it does when viewed from such a distance, must be something over 2,000 miles in diameter. This makes it a considerably smaller world than Earth, but one of respectable size. Its area is far greater than that portion of Earth's surface known to the Greeks.

The Moon was the *only* heavenly body whose distance and size could be worked out by observations with the unaided eye. The Greeks and those who followed them got no further than that. The Greeks tried to determine the distance, and therefore the size, of the Sun by trigonometric methods involving the Moon, but the measurements required were so tiny that the unaided eye simply did not suffice, and their estimate was far too low.

As for the other objects in the sky, the remaining planets and the stars, the Greeks could not even guess.

If the Moon had not been in the sky, there would have been no way of estimating the distance and size of any heavenly body prior to the invention of the telescope. There would have been no reason at all for supposing that there was a huge universe out there; that the sky and everything in it was particularly far away; or that the Sun and stars were more than they seemed to be, a small circle of light and tiny dots of light.

The Moon's presence changed that, however. It was a huge world hanging in the sky, and not so far away that it need be considered impossible to reach. It is not surprising, then, that there were stories of space flight, and specifically of flights to the Moon, even in ancient times

In the 2nd century A.D., about three hundred years after Hipparchus' determination of the Moon's size and distance, a Syrian satirist, Lucian of Samosata, wrote a tale (sardonically named *True History*) that is the earliest important piece of interplanetary science fiction surviving today. In this story, the hero sails out into the Atlantic Ocean and there is caught in a waterspout that carries him upward to the Moon.

The mere fact that the Moon was clearly a world roused suspicion about the other heavenly bodies as well. Might not the Sun be a world, too, despite its brilliance? Might not the various planets be worlds, and seem mere spots of light only because of their distance? In fact, Lucian of Samosata, in his romance, speaks not only of inhabitants of the Moon, but of their war with inhabitants of the Sun. His hero stops at the planet Venus, too.

Of course, Lucian took very little account of the difficulties in the way of reaching the Moon. The distance (which was enormous by Earthly standards of the day) was slurred over, as were the unknown, but clearly still greater, distances to the other heavenly bodies. Nor was there any feeling that the environment between Earth and Moon, or that of the Moon's surface, might be hostile to life. It was generally assumed that the air filled all of space out to the Moon, and that the Moon was as comfortable a world as the Earth.

After all, there was no reason to think otherwise in ancient times.

The decay of ancient learning in the later centuries of the Roman Empire was followed by a period of a thousand years or so in western Europe in which the scientific findings of the Greeks were forgotten, and in which a Biblical view of the Universe prevailed. The heavenly bodies were only lights once more, in line with the Biblical passage: "And God said, Let there be lights in the firmament of the heaven to divide the day from the night; and let them be for signs, and for seasons, and for days, and years: And let them be for lights in the firmament of the heaven to give light upon the earth: and it was so. (Genesis 1:14-15.)

Beginning in the 11th century, the science of the Greeks, preserved through the centuries by the Arabs, began to penetrate the darkness of western Europe once more, and learned men once again began to view the Moon as a world and the Universe as large.

In the 15th century, the mental horizons of Europeans began to stretch outward in line with the new age of exploration that was beginning. Mariners had the compass and were willing to venture out into the ocean, to find islands in the Atlantic, and to work their way southward along the shores of Africa.

And if Earth was larger than medieval man had dreamed, might not space be larger as well? In 1440, Nicholas of Cusa, a German cardinal, learned in Greek views, went further than any of the ancient philosophers had. He published a book in which he maintained that the Earth rotated on its axis; that space was infinite; that the stars were other suns and bore in their grip other worlds like the Earth and the Moon.

It was a daring concept and quite modern in flavor, but even the most daring flights of imagination must be based on some seed of fact, and for Nicholas's magnificent picture of the Universe it was the fact that the Moon *was* a world—and if there could be one world other than Earth, why not an infinite number?

Then came Christopher Columbus's discovery of the American continents in 1492 (a generation after the death of Nicholas of Cusa) and in the most dramatic way possible it was shown that the ancient Greeks had not known everything—deep as their thought had gone, it had not gone deep enough. European scientists became more adventurous in their attempts to improve on the Greeks.

In 1543, the Polish astronomer Nicholas Copernicus worked out the mechanics of a Solar system in which the planets circled the Sun and not the Earth. His system explained the apparent movements of the planets in the sky much more clearly and efficiently than did the old Earth-centered notions of the Greek astronomers.

The Copernican theory removed Earth from its unique position at the center of the Universe and made it merely one planet of many. If the Earth was a world, there was no reason to suppose that the other planets were not worlds too.

But even by Copernican notions, the Moon remained a companion world of the Earth. Although all the other planets, even Earth itself, might circle the Sun, the faithful Moon still circled the Earth. There might be many worlds, but the Moon was the closest, the most intimately related, and the logical first target for the men of Earth.

23

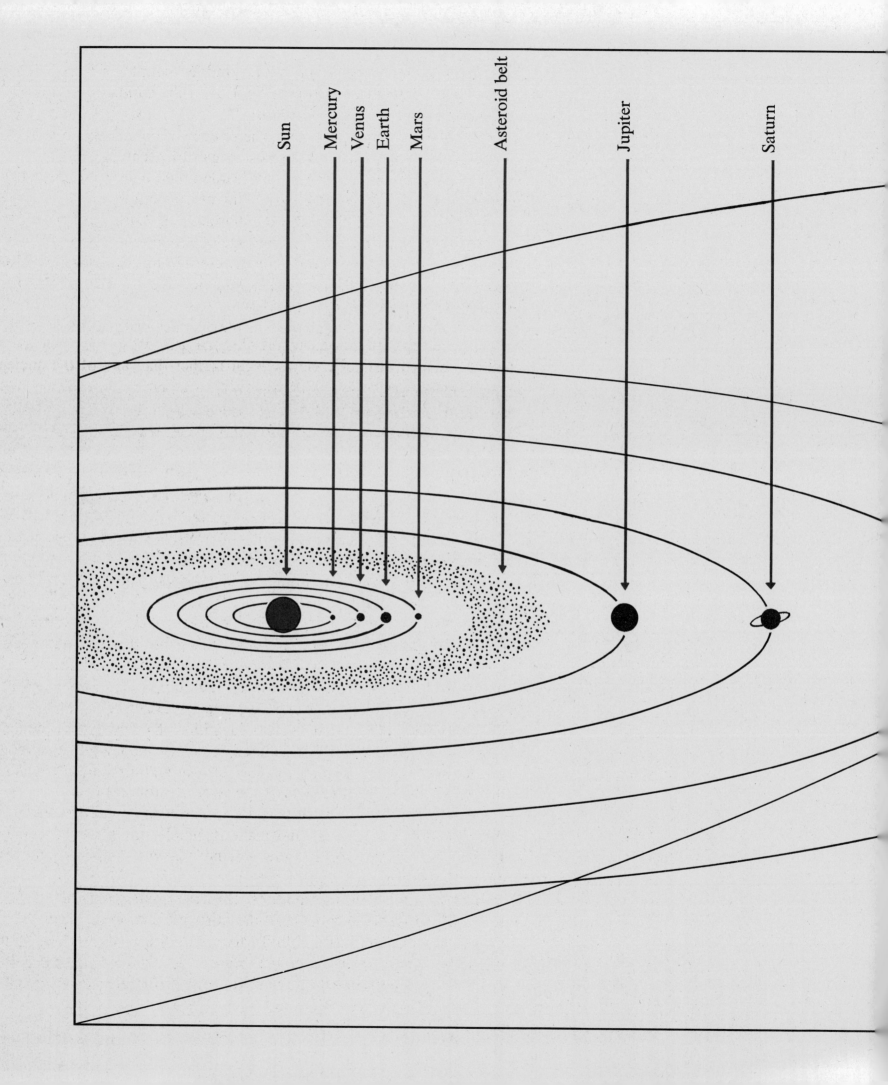

Sun Mercury Venus Earth Mars Asteroid belt Jupiter Saturn

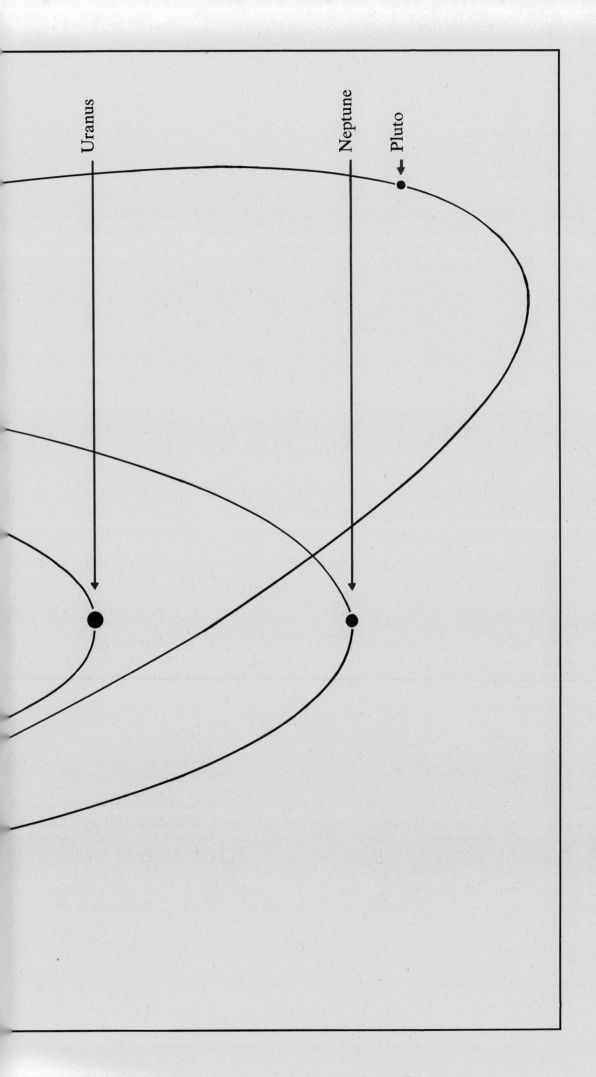

Uranus

Neptune

Pluto

The Solar System and
Orbits of the Planets

**Relative sizes and distances of the planets
from the Sun are indicated approximately.**

The great turning point came with the invention of the telescope. According to legend, the telescope was invented in the Netherlands in 1608, through the idle habits of a spectacle-maker's apprentice. This lad, wasting his time playing with lenses, noticed that when he held one type of lens a certain distance before one of another type, he could see distant things more clearly, and seemingly enlarged. His master, Hans Lippershey, seized on this finding by fixing the appropriate lenses in a metal tube—and had the first telescope.

The Netherlands was at war with Spain at the time and an attempt was made to keep the telescope secret, as a war weapon. Rumors of its existence reached Italy, however, and the Italian scientist Galileo Galilei quickly produced a telescope of his own in 1609.

Galileo did something the Dutch had not done. He turned his telescope on the heavens—and quickly made a host of discoveries. He found, for instance, that there was a myriad of stars that were not visible to the unaided eye. He found that the planets appeared as small globes in the telescope, supporting the contention that they were worlds that only seemed to be dots of light because of their great distance. He discovered tiny bodies circling Jupiter, making it clear that some heavenly objects definitely did *not* circle Earth and thus made the Copernican theory more plausible by that very fact. He noted that Venus showed phases like the Moon, which meant that Venus, like Earth and the Moon, and therefore probably all the planets, shone by reflected light only.

All these discoveries, however, supported the notion of other worlds in space only indirectly. Even if some of the heavenly bodies were large enough to show as globes (the stars remained mere dots of light even in the telescope), they might be merely featureless spheres. For all one could see in the first telescope, these bodies still seemed nothing at all like Earth, except in size and motion and in the fact of shining by reflected light.

But Galileo's very first telescopic observation gave evidence that this was not so, for when he lifted his telescope to the heavens, he looked first, of course, at the Moon. What he saw was a world on whose surface there were mountains. He also saw flat regions which he thought might be seas. Thanks to the telescope, it could finally be seen—not by subtle deduction, but by the direct evidence of one's eyes—that the Moon was a world with a rough, uneven surface like that of the Earth.

It is not surprising that the mind of man began, more than ever now, to contemplate the possibilities of a voyage to the Moon. Men had ventured into the wilderness of a wide ocean with no known shore in order to find land that *might* be there. Columbus had ventured

26

on a seven-week journey into the wastes to discover America. The Moon was eighty times as far from Europe as America was, but at least the Moon was in plain sight.

Thus an English clergyman, Francis Godwin, wrote a story called *Man in the Moone,* which was published in 1638. Godwin's hero (a Spaniard, and fittingly so, since Spaniards were among the great explorers of the time) flew to the Moon in a chariot hitched to great geese which were supposed to migrate to the Moon regularly. Godwin described the Moon as a land much like the Earth, but better. There was no attempt whatever to describe the Moon's environment in accord with what astronomers of his own time knew. To Godwin, the Moon was merely a new America.

This policy continued for two centuries, even though astronomers learned more and more about the Moon. Trips to our neighbor world remained mere adventure tales, with the Moon the convenient peg for those adventures. It might as well have been a mysterious island in the South Seas, or a land beneath the sea or under the earth, or a totally imaginary place outside our ordinary Universe altogether. (And indeed all such places, and many more, were used for adventure tales of science fiction.) Even as late as 1835, the *New York Sun* was able to run a series of stories about a Moon on which there were Earthlike conditions and intelligent life, and find that the general public believed it. The *Sun's* circulation figures boomed.

Oddly enough, even before Godwin's story, a science-fiction story had appeared that *did* take into account the actual astronomic situation. The author was Johannes Kepler, the great German astronomer, who had demonstrated in 1609 that the orbits of the planets were not circular (as the Greeks, and even Copernicus, had thought), but elliptical. It was Kepler who worked out the structure of the Solar system that we accept to this day.

Kepler also gave a name to the small bodies that Galileo reported circling Jupiter. Kepler called them "satellites" after a term used by the Romans for hangers-on who fawned on rich men and constantly circled them in the hope of being invited to dinner. And if those newly discovered worlds were satellites of Jupiter, then it was plain that the Moon was a satellite of Earth.

Kepler wrote a story about a flight to Earth's satellite. It was called *Somnium* and was published in 1634, a year after his death. In the story, the hero is transported to the Moon in a dream; but if his transportation was fantastic, the conditions he found on the Moon were real—given the knowledge of the time. For instance, in Kepler's story the day and night on the Moon are each two weeks long, as they are in reality. Because of this attempt to take scientific knowledge into account, Kepler's story might be considered the first true science-fiction tale, as opposed to mere fantasy.

The advance of science in the wake of the invention of the telescope rapidly lessened the attractiveness of the Moon as a world.

In 1643, the Italian physicist Evangelista Torricelli measured the pressure of a column of air by balancing it against a column of mercury. He found that the pressure of air would support a 30-inch-high column of mercury and thus invented the first barometer—and also provided information important for space exploration. To produce a pressure balancing the pressure of 30 inches of mercury, the Earth's air would have to extend five miles high, assuming it to be the same density on high as it is at the surface. In actual fact, the density of the air decreases with height so that the air layer extends far higher. Even a hundred miles above the surface of the Earth rarefied wisps of air can be detected.

However, air dense enough to support flying birds or to support the breathing requirements of animal life is confined to within a few miles of Earth. The 240,000 miles separating Earth from the Moon is (except for the five miles or so nearest the Earth's surface) essentially airless. This means that man cannot reach the Moon by the lifting effect of waterspouts or by the efforts of flying birds.

28 As late as the 1830's, the American writer Edgar Allan Poe wrote

a not-very-serious story entitled *The Unparalleled Adventure of One Hans Pfaal* in which the hero, Hans Pfaal, reached the Moon by balloon. This took advantage of a new invention, the balloon, which was first used to lift men into the air in 1783. Poe was careful to say that the gas used in his story was much lighter than hydrogen used in the balloons of the time, and therefore had more lifting power. But in actual fact no balloon, however light the gas it used, could possibly lift anything beyond Earth's atmosphere or get even a ten-thousandth of the way to the Moon.

Oddly enough, the first suggestion of the one possible means of reaching the Moon through the vacuum beyond the Earth's atmosphere came not from any man of science, but from a science-fiction writer—none other than Cyrano de Bergerac. Cyrano de Bergerac, the long-nosed duellist, was not a creature of fiction. He really existed, really had a long nose, really fought duels—and was also a clever writer. In 1650, he published a book called *Voyage to the Moon* in which he suggested a number of methods of reaching the Moon. He was apparently not very serious, though all his ways are ingenious. One method he mentioned was tying skyrockets to his chariot and zooming off by rocket power.

And that's it. The rocket engine is the one propulsive method known to man by which it is possible to cross a stretch of vacuum and

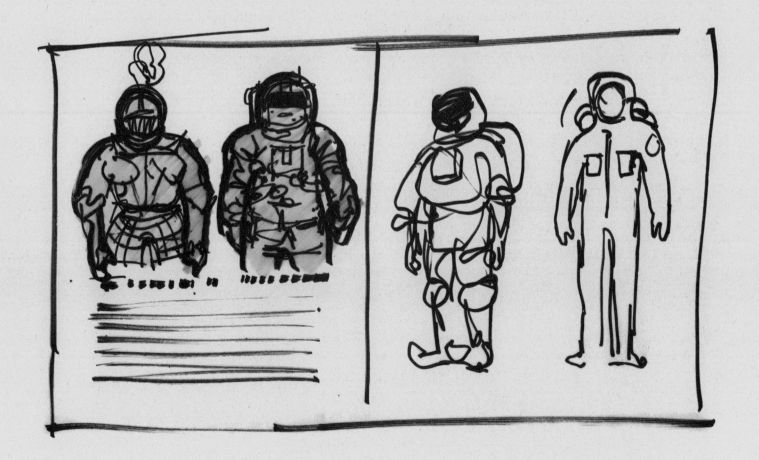

steer a course through space. It is the method by which mankind finally (three centuries after Cyrano's tale) reached the Moon.

Though Cyrano hit on the basic method of propulsion across the vacuum between Earth and Moon, he was not yet thinking about it, for he made no effort to protect his hero against that vacuum. But within a few decades Newton's work would make it apparent that the protection would have to extend to the surface of the Moon itself.

The possession of an atmosphere is not unique to Earth. The planets Venus and Jupiter clearly have atmospheres, as even a cursory telescopic study will show.

The Moon, however, is another thing. Its surface features, its mountains and craters, are seen always with perfect clarity and are never obscured by clouds or by even the tiniest trace of mist. (*Earthly* clouds and mist will, of course, easily hide the whole satellite, and often do.)

So in the decades after Galileo's discovery it became quickly apparent that the Moon lacked an atmosphere. And if it lacked an atmosphere, any water present would boil away quickly in the Sun's rays. Consequently, there ought to be no surface water either on the Moon. Closer investigation of the Moon's "seas" showed that they were in fact waterless. (Nevertheless, they are still called "seas" to this day, and even retain the romantic names given them in the 17th century, such as "Ocean of Storms," "Sea of Dew," "Sea of Foam," and so on—though the names are in Latin so that they seem less incongruous to English-speaking people.)

In 1687, the English scientist Isaac Newton published a book called *Mathematical Principles of Natural Philosophy,* which is probably the greatest single production in history by a scientist. In it, Newton worked out the laws of motion and the theory of universal gravitation. In effect, he demonstrated the working machinery of the Universe.

For one thing, he explained exactly how rockets would work, even in a vacuum, so that he put Cyrano's feat of imagination into the realm of the possible and natural. For another, he showed that the Moon's surface gravity had to be smaller than Earth's. (It is only one-sixth of Earth's, we now know.) The Moon's gravity was simply not strong enough to hold an atmosphere like Earth's, or water vapor either. After Newton, it was quite clear that the Moon could not possibly have either air or water (despite those science-fiction stories that continued, for centuries, to give our satellite both).

The Moon was now exposed as a harsh world, hostile to life: no air, no water, long days and nights of extreme heat and cold respectively, and a low gravity to which man's body is unused.

The first person to write science fiction in which all such matters were of serious concern was Jules Verne, working in the latter half of the 19th century. He was not a scientist himself, but he was the first science-fiction writer to research his work carefully. What is more, he was the first science-fiction writer of any kind, in the sense that he was the first writer to make science fiction his stock in trade and to depend upon it as the backbone of his literary income.

Verne wrote what was far-and-away the most popular tale of flights-to-the-Moon that had yet appeared, and one that made the greatest impression on the general public. Interestingly enough, in view of what was to happen in the future, Verne was particularly popular in Russia and in the United States.

His story, *From the Earth to the Moon,* was published in 1865. In it, Verne has the Americans carrying through a flight to our satellite, and thus he was the first to suggest that the first men to do so would not be Europeans. This made sense, since the Americans had just fought their bloody Civil War and were, for the time, expert in the use of artillery. It was by artillery Verne planned to reach the Moon.

Verne pictured Americans sinking a long, long cannon into the Earth. This would shoot a projectile at speeds great enough (seven miles per second and up) to pull away from Earth's gravity and reach the Moon. This is, indeed, quite possible, and an explosion powerful enough could blow a projectile to the Moon. However, the accelerations would be so great that any living thing inside the projectile would be smashed to jelly in the first seconds after the explosion. A rocket, on the other hand, attains the required speed gradually, not in one explosive burst, and therefore mounts with small accelerations that men can endure.

Despite the retreat from rocketry to artillery, Verne was in other ways scientifically aware, careful in his description of the mechanics of the flight and what the men on board the projectile experienced. Oddly enough, he even had the first flight toward the Moon take place from Florida, as it eventually did. Verne's explorers did not land on the Moon, but merely circled it and returned to the Earth, so that he was spared the necessity of describing the Moon's surface and the methods by which men could survive on it.

Verne's success brought a host of imitators and the story of space flight became common. The best of the post-Verne writers was the Englishman Herbert George Wells, who wrote *The First Men in the Moon* in 1901.

In Wells's novel, the means of propulsion was a metal that served as a gravity-insulator. A sheet of such metal, impervious to gravity, would naturally rise up into the air without power and would carry anything on it upward too. A spaceship built of such a metal would go to the Moon without trouble.

Actually, this is an even greater regression than Verne's giant cannon. The cannon is possible, though not practical. Wells's gravity-insulator is flatly impossible according to modern theories of the nature of gravity.

Wells, however, had his men land on the Moon and there they discovered a world that rather resembled that of Kepler's *Somnium*.

Science fiction continued to expand after Wells and the 20th century has seen it become a recognized and respected major branch of fiction (though the term "science fiction" was not actually invented until 1930). Although space flight and space exploration was by no means the only kind of plot in science fiction, it remained the favorite. In the 1930's (when there were popular magazines devoted to science fiction—the first of them, *Amazing Stories,* had appeared in 1926) writers grew tired of merely reaching the Moon and went joyfully off into the greater depths of space. But it was always the Moon adventure they could tell in the greatest and most accurate detail. After all, so much was known about the Moon.

So it was that the astronauts who finally stood upon the Moon seemed very much like the imaginative drawings that had filled the science-fiction magazines for forty years.

The fictions of Verne and Wells and their successors did more than merely imagine a future. They must often have acted as direct stimulants to imaginative engineers who turned their attention to the problem of exploring space and reaching the Moon, and who tried to solve the actual technical problems involved.

1 Where We Stand

The lunar lander Eagle touches down on the surface of the Moon: the first landing. A cutaway view showing astronauts Armstrong and Aldrin at the controls, and to the left the earlier stages in the operation —the lunar lander separating from the command module and the lander descending.

preceding page: The first men on the Moon: Neil Armstrong and Edwin E. Aldrin, Jr., July 20, 1969. (Collection Broadway National Bank, San Antonio, Texas)

this page: E.V.A. on the Moon. The first landing was followed by five other successful missions (as well as by the ill-fated Apollo 13). These scenes show the extravehicular activities of Apollo 17 mission commander Eugene Cernan and geologist astronaut Harrison Schmitt.

Skylab—frustration, suspense, and Yankee ingenuity. The first orbiting space laboratory was scheduled for an 8-month mission, manned successively by three crews of three. One of its primary goals was to test the effects on the human body of prolonged living and working in a weightless environment. During Skylab's launch on May 14, 1973, a thermal shield was torn away, taking with it one of the main solar panels and jamming the other. The resulting scorching temperatures inside Skylab and severe loss of power seemed to doom the mission, but a series of inventive—if dangerous—operations saved it. Astronauts Conrad, Kerwin, and Weitz erected an exterior sunshade to cool the ship and later Conrad and Kerwin took a space walk to free the jammed solar panel. The painting shows Skylab after this rescue operation.

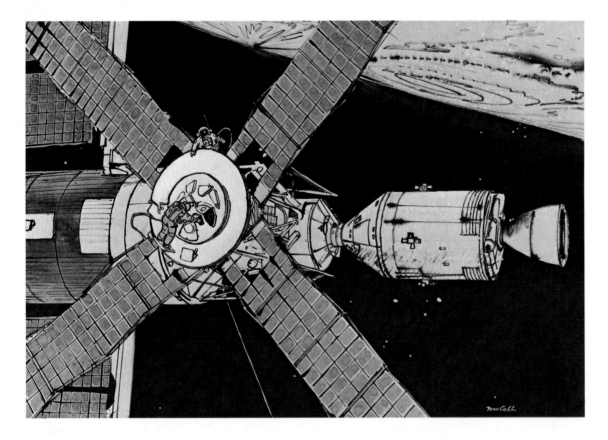

The command module that brought the crew up to the unmanned lab is shown docked to its nose. The four windmill-like solar wings are mounted on Skylab's complex telescope, which was used to make extensive, and extraordinarily significant, observations of the Sun.

The astronauts float about their tasks inside Skylab. In addition to the all-important biomedical experiments, the crew undertook extensive Earth resource studies and mapping operations, and conducted tests on metals in weightlessness that might someday lead to space manufacture.

Skylab crewman testing astronaut maneuvering unit. (The unit was tested inside Skylab, although it is intended for exterior use; courtesy of artist's license, it is shown that way here.) At right is the Apollo telescope mount surrounded by the four solar wings that were the principal source of solar power for Skylab before the crippled panel was deployed.

2 The Rocket

It is all very well to dream of going to the Moon, but how does one actually do it?

The difficulty is such that, in fiction, dreaming is accepted as a literal method of transportation. Kepler's hero in *Somnium* went to the Moon in a dream. In Edgar Rice Burroughs' first romance laid on the planet Mars, his hero, John Carter, gets there by desperately wishing to be there.

Dreams aside, however, the trip to the Moon must begin by rising, somehow, from the surface of Earth; and man, unaided, cannot do even that, except momentarily for the few feet a jump will allow him.

Before man's eyes, however, is the example of the bird, which can fly at will; some can even attain great heights in the air. Birds are too small to be useful to man, but imagine a horse with wings—the Greeks did, and called it Pegasus.

Or suppose men built artificial wings, attached them to their arms, and then flapped themselves into the sky, bird-fashion—as in the Greek myth of Daedalus and Icarus. (But Icarus flew too near the Sun, and the wax that held the feathers to the wing-frame melted. Off went the feathers, and down went Icarus to his death.)

But the example of the bird was fruitless. Flying creatures are upheld by the air and, in view of the low density of air and the high gravity of Earth, the wing surface must be comparatively large to support the weight of the body. The larger the body, the greater the wing surface, at an ever-increasing ratio. Above a rather small weight, the muscle-and-wing machinery required is more than evolution has been able to bring about. It is doubtful if any creature weighing more than fifty pounds has ever been able to fly by its own muscle power. The heaviest flying bird, the Kori bustard, weighs no more than this. The largest flying creature in the Earth's history, the long-extinct Pteranodon, had a wing-span of 27 feet, but it was almost all wings and probably weighed no more than 40 pounds.

For a man, then, even a small man, to fly under his own power, is quite out of the question. We must assume that if man is to rise from the ground for any extended period of time and attain any consider-able height in guided flight (as opposed to merely floating with the wind as in a balloon), some force greater than that available to animal muscle must be involved.

One inanimate force clearly capable of lifting heavy objects into the air, objects far heavier than birds, is the wind. The effect of hurricane winds is familiar even to those of us who have not experienced them directly, and a tornado can lift, or destroy, or both, anything that is not an integral part of the Earth.

Naturally, then, there have been stories of people rising into the sky in a whirlwind or tornado. In the Bible we have Elijah who "went up by a whirlwind into heaven." (II Kings 2:11). Another case, perhaps even more familiar to us, is that of Dorothy, who reached the Land of Oz by way of a tornado.

There is a sharp limit, however, to the lifting ability of a tornado, even if we disregard the damage it would do to anything it carried. The tornado can rise no higher than the lower reaches of the Earth's atmosphere, and, for that matter, neither can a balloon or any living thing. No force involving the air, whether a tornado or a body supported by air, can rise higher than a few miles.

If we are to reach the Moon at all, we must have some force that is not only strong enough but also capable of working in a vacuum.

The force that was finally to succeed was shadowed forth by gunpowder, which was originally discovered in China. How it was discovered, exactly when and by whom, is unknown, but the key factor of that discovery was the use of saltpeter, more formally called potassium nitrate.

Gunpowder is a mixture of potassium nitrate with charcoal (carbon) and sulfur. Both carbon and sulfur are known to burn, but they do so with relative slowness, through combination with the oxygen in the air. Potassium nitrate contains oxygen atoms as part of its molecule and these oxygen atoms can be broken away with relative ease. The result is that when potassium nitrate is mixed intimately with carbon and sulfur, and the mixture is heated to the point where the oxygen atoms begin to break away, all the carbon and sulfur combine rapidly with the oxygen atoms made available in all parts of the mixture.

If gunpowder is heated in the open, there is a flash and a great deal of gas is evolved. If gunpowder is confined, the gas evolved creates a huge pressure that is likely to break the container, producing the sound and effect of an explosion.

The explosive force of gunpowder can be used in various ways. Suppose you start with a cylinder open at one end. A quantity of gunpowder is pushed against the closed end of the cylinder and an object is pushed down the cylinder to confine the gunpowder. When the gunpowder is ignited, the pressure of the gases pushes the confining object out of the cylinder with great force. This, essentially, is what we have when a cannon fires a cannonball.

Alternatively, the gunpowder can be placed in a cylinder with a narrow opening at one end. When the gunpowder is ignited, the gases escape forcefully through the narrow opening and the entire cylinder moves in the opposite direction. This, essentially, is a rocket.

Actually, neither device is as easy as it sounds. For a cannon to work, the cylinder must be strong enough to withstand the explosion, and the bore must be straight and even enough to allow the cannonball to move easily up and down it, but narrow enough so that the gases cannot escape around the ball and weaken the explosive force.

For a rocket to work, the gunpowder must be prepared to burn slowly and evenly so that the gases can escape through the constricted opening without piling up to the point of explosion.

On the whole, it is easier to make rockets than cannons with relatively fragile materials, and the first record we have of the use of gunpowder in warfare, during the siege of a Chinese city by Mongols in 1232 A.D., was in rockets. The besieged used "arrows of flying fire," undoubtedly rockets. They had only psychological effect, however, annoying the Mongols and startling their horses. The Chinese lost in the end.

In the following decade, the Mongols built a huge empire stretching from the Pacific Ocean to central Europe, and it is not surprising that news of gunpowder spread westward with them. By 1280, western scholars such as Roger Bacon of England and Albertus Magnus of Germany were talking about gunpowder.

In western Europe, it was the catapult action of gunpowder that attracted men: Instead of hurling huge stones by the lever action of ordinary catapults, hurl them by exploding gunpowder. In 1324, simple cannons were used by the citizens of the city of Ghent, and on August 26, 1346, they were used by Edward III of England at the Battle of Crécy. It was a great victory for Edward, but not because of the cannon, which were still too small to do more than frighten horses. Nevertheless, cannon continued to improve, and by the mid-15th century, they had destroyed the basis for medieval warfare and put an end to feudalism.

Cannon could do more damage than rockets in those early centuries because it was only the cannonball that moved. The stationary cannon could be made massive and strong, so that in the end large cannonballs, moving at great speed, could batter down castles and city walls. In the case of rockets, the entire cylinder had to move, so they could not be made very massive. Furthermore, a rocket could be used only once, while a cannon could be used over and over. On the whole, then, rockets were not used in war but were retained in the guise of fireworks, to amuse and startle spectators; they were not forgotten.

Gunpowder is important in that it does not require air either to burn or to support the objects it sends into air. It contains its own oxygen supply in the potassium nitrate, so that it can explode in a vacuum as easily as in air. Furthermore, a cannonball sent up into the air rises because of the original impetus of the explosion and not because of the support of the air, unlike a balloon or a bird. In fact, the resistance of air slows the cannonball, and it could rise more efficiently in a vacuum than in the air.

To be sure, as a cannonball rises, it is under the constant pull of Earth's gravity, and therefore its speed gradually decreases. Finally it comes to a halt and begins to fall. Naturally, the greater the force of the explosion, the more rapidly the projectile moves to begin with and the longer before its speed is reduced to zero by gravitational pull, and the higher it will rise. Furthermore, the higher it rises, the weaker the effect of gravitational force becomes, since this force decreases as the distance from the Earth's center increases.

If the initial velocity is great enough, the ball will rise to great heights so quickly that the rapidly weakening gravitational force of Earth will *never* suffice to bring its speed to zero. It will escape from Earth altogether and take up an independent orbit about the Sun. (If it moves fast enough to begin with, it will escape from the Sun's pull, too, and move off into interstellar space.)

The minimum initial speed required to escape Earth's pull is 7.0 miles per second. If a cannon could fire an object at this initial speed, and if it were correctly aimed, that object would reach the Moon (as in Verne's *From the Earth to the Moon*).

The objection to this method of transporting delicate devices, and especially *people*, to the Moon, is that the spaceship must be brought from a speed of zero to the speed of 7.0 miles per second in the time it takes to travel through the bore of the cannon. Such an acceleration could not possibly be endured. What is more, the rapid motion through the atmosphere after the object leaves the cannon would result in friction that would probably melt the object and fry anything living that was inside.

Neither objection applies to the rocket. The rocket rises even while the gunpowder is burning, and the acceleration continues until the gunpowder is completely consumed. If enough gunpowder is used, the rocket will not have to reach the necessary velocity required to pull free of Earth's gravitational pull until a height of many miles is attained, and the necessary acceleration, spread out over so long a period of time, will be low enough at any given moment for human beings to endure. Furthermore, by the time the rocket attains sufficient speed to escape, it is so high that the air about it is very thin and does not produce enough friction to raise its temperature dangerously.

The question is, though, will a rocket continue to work once it leaves the atmosphere—if, for instance, it is necessary to put on more speed, or to change direction, or to slow down?

Many people don't see how it can. We are accustomed to thinking of a motion upward as the result of a push downward. We jump up by pushing off the ground with our feet. Birds fly by pushing the air with their wings. Cannonballs rise because the exploding powder pushes against the closed bottom of the cannon cylinder.

Surely, then, the rocket, too, must push against something as it rises. The only thing it can push against is air, and if that is so, it cannot exert force in a vacuum; it is as helpless in space as a cannonball, and it must get all the speed it needs (and in the correct direction) before it leaves the atmosphere.

In 1687, however, Isaac Newton, in his great book, showed as his "third law of motion" that action and reaction are equal in amount and opposite in direction. That means that only objects that rise *entirely* must push against something. All of our body jumps while the Earth moves (very slightly) downward. All of a bird flies while air is pushed downward. All of a cannonball rises while the cannon is pushed downward.

But suppose *part* of an object at rest is hurled downward. In that case, the rest of the object must rise upward to balance it, by Newton's third law of motion. In a rocket, part of the substance (the gases arising from the burning gunpowder) is hurtled downward. The rest of the rocket is therefore hurtled upward. This does not depend upon a push against air or against anything else, so that a rocket can operate as easily in vacuum as in air. In fact, it can operate far more easily in vacuum for there is no air to present resistance and slow both the downward exhaust and the upward rest-of-the-rocket.

What Newton showed in theory has been since amply demonstrated in practice. The rocket was the first object created by man that could: a) move through a vacuum; b) accelerate slowly enough to make it conceivable that men could survive; c) attain speeds high enough to enable it to reach the Moon, and reach those speeds high enough in the atmosphere so that air resistance would not heat it dangerously.

In the three centuries that have passed since Newton's time, mankind has discovered nothing other than the rocket principle that could make manned space flight possible. The rocket remains the one and only. 45

To reach the Moon, of course, a rocket has to attain extraordinary speed, and no rocket built prior to the 20th century could develop the necessary speed. If rockets had remained solely an amusing fireworks device, it might never have occurred to anyone to try to design one that could attain that speed. In Napoleonic times, however, rockets entered warfare again in such a way that they gave rise to serious thoughts once more.

This came about in the latter half of the 18th century, when the British embarking on their conquest of India encountered the use of rockets by the Indians. These were not the cardboard rockets used by the Chinese many centuries before, or by Europeans in their fireworks displays. These were made of iron tubes, and weighed six to twelve pounds. When they came whishing at a line of soldiers, they did more than frighten horses.

A British officer, William Congreve, persuaded his superiors to build rockets of their own. These were used on several occasions during the Napoleonic wars. They carried incendiary matter, and Copenhagen was severely burned by a British rocket barrage in 1807. By the end of the Napoleonic wars, the British were using rockets with a range of over half a mile. (They were used at the siege of Fort McHenry, near Baltimore, on the night of September 13-14, 1814; it was on that occasion that Francis Scott Key wrote "The Star-Spangled Banner," which contains the reference to "the rockets' red glare.")

Rockets did not continue to gain importance in warfare, however, because ordinary artillery improved and its performance rapidly outstripped that of rockets once again. Still, Congreve had made rockets spectacular enough so that there was speculation about using rockets as a means of propelling objects. It was, after all, a time when new means of propulsion were coming into use. The steam engine used the pressure of the vapors produced by boiling water to turn wheels. By 1807, such steam-powered wheels were used to propel ships through water, and by mid-century steam locomotives made the railroad train an important factor in transport.

Travel through the air also became possible when the first man-carrying balloons rose in 1783. There was no reason the balloons could not be made to do more than float helplessly with the wind. It would only be necessary to place a steam engine in the gondola and have it turn a propellor. Of course, building a balloon big enough and sturdy enough to carry a heavy steam engine and its water and fuel supply was a difficult job and it was not accomplished until 1900—but the principle was understood long before.

The use of steam power and then later of internal-combustion engines, involved pushes against the outside world. Steamships and steam loco-motives and automobiles and propeller airplanes move forward all-in-a-piece. For that to happen, something outside themselves must move backward: water, ground, or air. To be sure, vapor escapes from any steam engine and exhaust from any internal-combustion engine, but it does so slowly and in small quantity and has no real effect on the motion of the vehicle.

The rocket principle could be used to propel objects across land, over water, or through the air, but always that motion would be less efficient, harder to control, and more expensive than motion-by-external-push. Through the 19th century, inventors kept trying to harness the rocket for ordinary land, sea, or air travel—and kept failing. Only when it came to travel beyond the air, out in space, did the rocket principle come into its own. Out there, only the rocket principle would work.

The first person to think of space flight in terms of rockets, and to do so with due attention to scientific principles, was Konstantin Eduardovich Tsiolkovsky, a Russian schoolteacher, born on September 17, 1857. He was largely self-educated and spent much time working out on his own certain scientific laws which he was unaware were already known.

In 1903, he began a series of articles for an aviation magazine in which he went into the theory of rocketry quite thoroughly. He wrote of man-made satellites, for instance.

These had been first suggested by Newton as a consequence of his theories of motion. If an object is falling toward the Earth and is at the same time moving at right angles to the direction of the Earth, it will fall to Earth a considerable distance from the spot directly underneath when its fall began. If it is moving fast enough, the curve of the Earth moves away from the falling object as fast as the object itself falls—and the object falls forever, so to speak, without reaching the Earth (provided it is outside the atmosphere so that air resistance does not slow it and start to curve its motion more sharply downward). The object, although in "free fall," moves around the Earth.

The Earth moves around the Sun in free fall, falling always toward the Sun, but moving sideways at the same time so that its curving fall matches the curve of the Sun's surface. The Moon moves around the Earth in free fall, and if we can send an object up above the Earth's atmosphere and give it a proper speed and direction, it can also move around the Earth, but at a distance from us much smaller than that of the Moon. The Moon is a natural satellite; the man-made object moving about the Earth would be an artificial satellite.

There were obvious advantages to such an artificial satellite for 47 the development of space flight. In the first place, the velocity required to place a satellite in orbit just outside Earth's atmosphere is only about 5/7 that required to allow it to escape Earth's gravitational pull. This means that a satellite need move only 5 miles per second, a speed that could be attained before it was practical to produce the 7-mile-per-second velocity required to reach the Moon.

Then, too, if such a satellite were large enough, it could carry men and become a "space station." Tsiolkovsky was the first to suggest the possibility of such a station.

Tsiolkovsky also recognized that gunpowder would not do for space-ships. Gunpowder could produce gases to propel the rocket; the longer the gunpowder burned, the longer the period of acceleration and the greater the final velocity of the spaceship. However, the longer you want the gunpowder to burn, the more gunpowder you must carry and the greater the weight that you must lift. In the end, you cannot reach escape velocity that way: what you gain by adding more gunpowder to make the burning last longer is cancelled by the greater weight of that additional gunpowder.

Tsiolkovsky therefore suggested something that would produce more thrust *per weight* of fuel. He decided it would be necessary to burn kerosene, and this was a move in the right direction.

A decade after Tsiolkovsky's speculations and calculations were published, an American physicist, Robert Hutchings Goddard (born in Worcester, Massachusetts, October 5, 1882), was independently going over the same ground. In 1919, he published a small book entitled *A Method of Reaching Extreme Altitudes,* which contained nothing that had not been explored by Tsiolkovsky.

Goddard, however, went beyond Tsiolkovsky in one all-important respect. Where Tsiolkovsky had been content to do nothing but calculate, speculate, and write, Goddard experimented. He actually built rockets and studied their performance.

At first he built gunpowder-powered rockets, but he quickly came to the conclusion, as Tsiolkovsky had, that gunpowder was not the answer. In 1923, he began to test engines that would use gasoline and liquid oxygen. The two liquids, kept in separate chambers, of course, would be fed into a common chamber at a controlled speed and there they would mix and burn. The rate of burning would be determined by the rate at which they flowed into the chamber.

In 1926, Goddard sent up the first liquid-fuel rocket, the rocket that was to be the direct ancestor of future space-flight devices. It was a tiny object, about four feet high and six inches in diameter, and was held up in a frame that looked like a child's jungle gym.

48

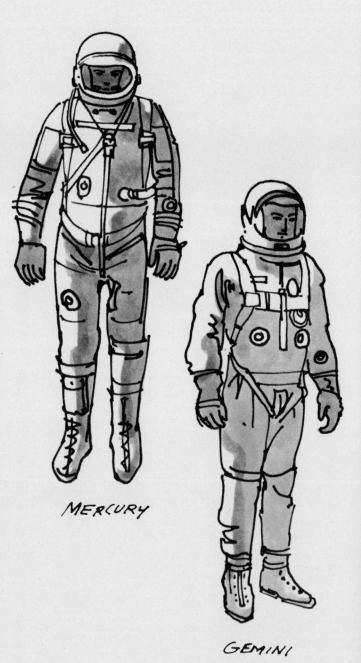

MERCURY

GEMINI

His wife took a picture of him standing next to it before the launch, and that picture is now the most familiar single photograph in the history of rocketry.

When that first liquid-fuel rocket was fired, it traveled 184 feet and reached a speed of 60 miles an hour.

Goddard continued to experiment, first in Massachusetts, then (after the neighbors complained and the law stepped in) in New Mexico. By 1935, he was firing rockets that went faster than the speed of sound and reached heights of a mile and a half. In the course of his experiments he designed efficient combustion chambers, devised cooling systems to prevent the walls of the combustion chambers from growing too hot, developed systems for steering a rocket in flight by using a rudderlike device to deflect the gaseous exhaust, with gyroscopes to keep the rocket headed in the proper direction. He accumulated a total of 214 patents.

His most important advance, perhaps, was to work out the concept of a multi-stage rocket. Why lift a huge rocket all at once with one supply of fuel? That way, after most of the fuel is expended and the rocket is high in the air, it must still lift the massive chambers that once held the full supply of fuel.

Suppose instead, there were three rockets, so to speak, one mounted on another. The bottom rocket would expend its fuel and lift the others some miles into the air. Then it would be detached, leaving only the two rockets above. The second rocket would now start burning its fuel, but it would do so when it was already miles above the surface, when it was already moving at a rapid speed, when it was surrounded only by thin air, and when it had to lift only itself and the third stage, not the huge, discarded first stage.

Unencumbered by the great weight of the first stage or by air resistance, the second rocket would increase its speed. Then, when it had expended its fuel, it would drop off, and the third rocket would carry on—from a still greater speed, with still less weight and air resistance.

The same amount of fuel used in three stages would move the final payload to much greater velocity and to a far greater height than it would if used in one stage. Of course, the engineering details required to release one stage and start the second add to the complexities of the rocket, but the advantages outweigh the problems.

Both Tsiolkovsky and Goddard were lone wolves. Each worked by himself and could do little in consequence. But in Germany the situation was different.

There, in 1923, a small book entitled *The Rocket into Interplanetary*

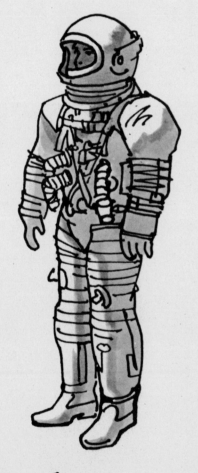

APOLLO

49

Space was published. Its author was Hermann Oberth, born in what is now Rumania on June 25, 1894. His book aroused enough interest to bring about the formation of a "Society for Space Travel" in Breslau, Germany, on June 4, 1927.

Naturally, this group could accomplish very little at first. It cooperated, however, in the 1928 filming of a science-fiction motion picture in Berlin called "The Girl in the Moon." This showed a rocket, and for the first time made use of a reverse countdown: *10, 9, 8, 7, 6, 5, 4, 3, 2, 1, fire*. The director, Fritz Lang, had thought this up for purely dramatic purposes.

By 1930, the organization had nearly a thousand members and was beginning to conduct experiments with liquid-fuel rockets. One of the new members of the society was a teen-ager named Wernher von Braun. (He was born in what is now Poland on March 23, 1912).

The experiments of the association were perhaps not as successful nor as impressive as those Goddard was conducting single-handed in the United States, but when the Nazi regime came to power in 1933, German rocket experiments were harnessed to the needs of the military and research intensified. Under the leadership of von Braun, German rocket research finally produced the "V-2," and rockets once again became a war weapon, in a form immensely more complex and dreadful than those used in the time of Napoleon.

The V-2 came into combat use in 1944, too late to win the war for the Nazis, although they fired 4,300 of them, of which 1,230 hit London, killing 2,511 and wounding 5,869. The V-2 had to be taken seriously. The United States might ignore the peaceful rockets of Goddard, but the war weapon of von Braun impressed American military men—and Soviet military men, too. Once Germany was occupied, the Americans and the Soviets raced for the V-2 rockets that remained unfired and for such rocket experts as could be rounded up.

By 1949, the United States had fired a captured German V-2 rocket to a height of 128 miles. In the same year, its rocket experts sent a WAC-Corporal, the second stage of a two-stage rocket, to a height of 250 miles.

Achieving such heights was, in itself, mere record-breaking, unless scientifically useful information concerning those heights could be obtained. Instruments might be placed on rockets (and had been by Goddard in his earliest experiments) and then recovered when the rocket returned to Earth. Safe return from great heights was uncertain, however, and from the greatest heights was not expected.

The answer was "telemetering" ("measuring at a distance"), first applied to atmospheric research by balloon in 1925 by a Russian

scientist, Pyotr A. Molchanoff. Essentially, this technique entails translating the conditions to be measured (anything from temperature recordings to television pictures) into electrical impulses of varying strength that are transmitted back to Earth as radio waves. These electrical impulses, varying in accord with whatever is being measured, are re-translated on Earth into the measurement originally taken by the instruments at a height.

Eventually telemetering became so elaborate that the rockets seemed to do everything but talk, and their intricate messages had to be interpreted by rapid computers. Indeed, the growth of electronic computers after 1944, and the invention of the transistor at the end of 1947 (which made it possible greatly to reduce the size of electronic instruments), were essential to attempts to reach the Moon. Only transistorized equipment made it possible for rockets to pack enough instruments into the payload to make the exploration of space useful; only transistorized computers could be compact enough and complex enough to make the necessary calculations that could guide the rockets in space.

American interest in space flight (as opposed to war-weapon research) remained mild until October 4, 1957. On that date the Soviet Union placed a satellite into orbit about the Earth, missing the centenary of Tsiolkovsky's birth by only three weeks.

The satellite, "Sputnik 1," did little more than circle the Earth every 90 minutes at a distance of from 156 to 560 miles above its surface, and signal its progress by a radio beep. It was, however, the harbinger of a benign competition between the Soviet Union and the United States, each attempting to score "firsts." As a result of this national rivalry, space exploration advanced further than it might have if either nation had been in the field alone.

On November 3, 1957, the Soviet Union orbited Sputnik 2, which contained a dog, the first living organism to be placed in orbit about the Earth—it died there. The first American satellite to be placed in orbit was Explorer 1 on February 1, 1958.

On January 2, 1959, the Soviet Union fired Luna 1, the first satellite to surpass Earth's escape velocity and to leave Earth forever. It was a "Moon-probe" and passed by the Moon at a distance of 3,700 miles from the surface. The first American Moon-probe was Pioneer 4, launched on March 3, 1959.

On September 12, 1959, Luna 2 hit the Moon, the first man-made object to find a resting place on another world. On October 4, 1959, Luna 3 passed around the Moon and sent back the first photographs of the hitherto hidden far side of the Moon. Then, on August 19, 1960, the Soviets followed these Moon "firsts" by sending two dogs into orbit in Sputnik 5 and recovering them safely after eighteen orbits —the first living organisms to return safely from orbital flight.

Meanwhile, American rocket engineers were launching a variety of communications satellites, weather satellites and navigational satellites. There were indications aplenty that space exploration meant more than just reaching the Moon. Earth's weather could be studied from space in ways that would not be possible by any other means. The possibilities of mapping the Earth, studying its resources, providing new and incredibly efficient methods of long-distance communication, all became realities.

In addition, a flood of information concerning the nature of space in Earth's neighborhood was received. Details of the upper atmosphere were reported. The existence of the "Van Allen Belts" (named for James A. Van Allen, the man who first interpreted the information sent back), a region with high concentrations of charged particles, was proved. The detailed shape of the Earth was worked out from small deviations in the satellites' orbits. The concentration of meteoric dust was studied, the Solar wind (high-speed charged particles shot out from the Sun in all directions) was measured, as were cosmic-ray particles.

It was, however, manned space flight that captured the world's imagination, and on April 12, 1961, Yuri Alexeyevich Gagarin in Vostok 1 became the first man to orbit the Earth. He made a single orbit in 108 minutes and was brought safely back to Earth. Later that same year, on August 6, Gherman Stepanovich Titov in Vostok 2 circled the Earth for seventeen orbits before coming safely back to Earth. Both were Soviet citizens.

The first American to go into orbit was John Herschel Glenn, Jr., who was launched on February 20, 1962, and circled the Earth three times before returning safely. Two other Americans were orbited that same year, while on August 11 and 12, the Soviets managed to put two manned ships into orbit at the same time.

In May 1963, the American astronaut L. Gordon Cooper, Jr., set an endurance record by remaining in orbit for nearly thirty-five hours and circling the Earth twenty-two and a half times. This was surpassed the next month when the Soviet cosmonaut Valery Bykovsky remained in space for 81 orbits, totaling 119 hours in space. At the same time, the Soviets placed Valentina Tereshkova in orbit; she was the first (and so far still the only) woman in space. In October 1964, the Soviets placed Voshkod 1 in orbit, with three men on board. It spent sixteen orbits in space and was the first multiply-manned space shot.

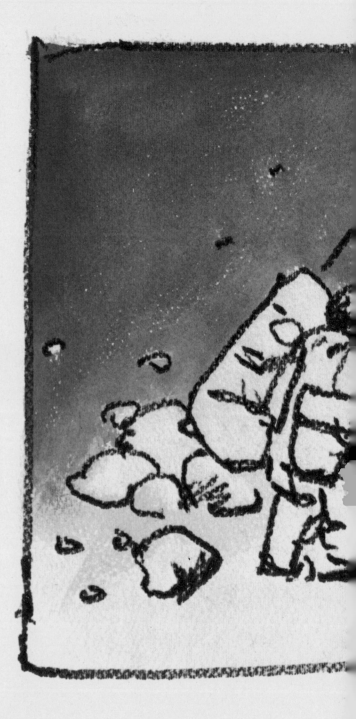

APOLLO 15 EVA

In 1964, the Americans took a clear lead in one aspect of space exploration when, on July 28, the unmanned Ranger 7 took more than 4,300 photographs en route to the Moon and sent them back to Earth before crashing into the Moon.

On March 18, 1965, the Soviets placed Voshkod 2 into orbit with two men on board. One of them, Aleksei Leonov, left the spaceship and spent ten minutes in open space, encased in a spacesuit and attached to the ship by a tether. It was the first "space walk."

That same year, the United States launched two-man spaceships whose orbits could be adjusted in space and which could be maneuvered from within. This was the first time any vessel in orbit was successfully maneuvered. In June, Edward White, one of the astronauts on Gemini 4, became the first American to walk in space. He remained in space for twenty-one minutes.

On August 21, 1965, Gemini 5 was launched with L. Gordon Cooper and Charles Conrad aboard. Cooper was the first man to be placed into orbit twice. Gemini 5 circled Earth for 128 orbits, remaining in space for eight days before coming safely back. Gemini 7, launched on December 4, 1965, orbited Earth 220 times and remained in space for two weeks.

On January 31, 1966, the Soviets launched Luna 9, which made the first soft-landing on the Moon. It took photographs of the Moon from its surface and sent them back to Earth. The Americans duplicated that feat with Surveyor 1, launched on May 30, 1966, which sent back over 11,000 photos.

Meanwhile, on March 16, 1966, the American ship Gemini 8 was launched with Neil A. Armstrong and David R. Scott aboard. It succeeded in docking with an unmanned vessel; this was the first docking in space, a maneuver that was essential to any manned landing on the Moon. More elaborate dockings were successfully concluded that same year.

APOLLO 15 EVA AUG. 171

Also in 1966, both the Soviet Union and the United States succeeded in placing unmanned satellites into orbit about the Moon. Luna 10 was launched by the Soviets on March 31, and Orbiter 1 by the United States on August 10. In this way, the entire Moon could be, and eventually would be, mapped in detail.

Until this time, the program for manned exploration of space had proceeded without human fatality. In 1967, the strokes of ill fortune, perhaps inevitable, came. In January, three American astronauts, Virgil I. Grissom, Edward H. White, and Roger Chaffee, died on the ground in a fire that broke out in their space capsule during routine tests. Grissom had been on Gemini 3, the first American rocket to carry more than one man. White had been the first American to walk in space. Later in 1967, on April 23, the Soviet cosmonaut Vladimir M. Komarov, who had led the three-man team that had flown the first multi-man rocket two-and-a-half years before, died when the parachutes of Soyuz 1 fouled on re-entry; Komarov was the first man to die in the course of a space flight. (Since then, three other Soviet cosmonauts died in space, in a three-man vessel.)

The accidents forced a delay in manned programs in both nations. The United States had been about to begin the Apollo program, a series of three-man launchings that would eventually bring men to the Moon. The first of these flights was delayed for a year and a half while the vessels were redesigned to provide additional safety against fire.

It was not until October 11, 1968, that the first flight of the new series, Apollo 7, was launched; it completed its mission successfully. Apollo 8, launched on December 21, 1968, flew out to the Moon and went into orbit around it before returning to Earth. Apollo 10, launched on May 18, 1969, did the same, and sank down to within nine miles of the Lunar surface in the process.

Finally, on July 16, 1969, Apollo 11 was launched, with Neil A. Armstrong, Edwin E. Aldrin, Jr., and Michael Collins on board. On July 20, while Collins piloted the main part of the vessel in orbit, Armstrong and Aldrin took a small lunar lander down to the Moon, and Armstrong became the first, Aldrin the second, human being to set foot on any world other than the Earth.

The dream of men from Lucian of Samosata down to the eager young science-fiction writers of the mid-20th century, was fulfilled. Of the important pioneers, only Wernher von Braun lived to see the day. Willy Ley, the German-American rocket expert who had been a guiding spirit of the rocket experiments in Germany, who had been instrumental in introducing von Braun to the world of rocketry, and who, through his writings, had done more than anyone else to keep rocket research before the eyes of the public, died six weeks before the Moon landing.

From the time the first feeble man-made satellite lifted into orbit, to the triumphant landing on the Moon, was not quite twelve years—twelve years that represented the most amazing saga of technological achievement for peaceful purposes (both in the United States and the Soviet Union) in the history of mankind. It was an achievement, moreover, in which the Moon landing was only a part (even if the most expensive part), for space exploration saw satellites of immense variety and function placed in orbit, and knowledge of all kinds added to the human reservoir.

The first astronauts to land on the Moon, after leaving instruments to make measurements in their absence, returned safely to Earth, bringing back samples of rocks from the Moon's surface.

Nor was it a one-shot. Five other trips, on Apollos 12, 14, 15, 16, and 17, have been successfully carried through. Each mission has performed more elaborate experiments and observations than the one before, all of it seen in clear television view from Earth. Men on Earth have seen men driving a powered vehicle on the Moon. Only one vessel aimed at the Moon did not make it. An accident in flight aborted the mission of Apollo 13, but the ship and its men were brought safely back to Earth.

By the end of 1972, twelve men, all Americans, had walked on the surface of the Moon.

2 The Next Steps

Many of the next steps have already gone beyond the drawing board. The space shuttle, the next major venture, is being readied for testing even as it undergoes continued modification. *Below:* Launch control at Cape Kennedy in the early 1980's: watching a space shuttle lift-off. The design concept pictured here, single large booster with piggy-back orbiter, was one of many considered by NASA.

The shuttle will combine the advantages of airplane and spacecraft, flying into space and returning to Earth to make future flights, resulting in enormous savings of money and time. *Right:* Kennedy Spaceport in the late 1980's. At right is an advanced shuttle system ready for launch, the small orbiter on the back of the giant manned booster. Another shuttle is being moved up to its launching pad. The orbiter will be essentially a common carrier, its passengers no longer required to pass stringent standards for space flight.

opposite: The orbiter as it separates from the booster—in this advanced concept the booster is manned also and will return to its launch base to be used over and over again.

this page, below: Flight profile of the shuttle system. The shuttle takes off vertically at left; its two solid-propellant boosters are jettisoned at a height of about 25 miles and descend by parachute to the ocean, where they will be recovered. The orbiter flies on, powered by the liquid propellant in its huge external tank. Once in orbit, the tank separates and is sent back into the Earth's atmosphere by a small rocket; it falls—already breaking up—into a remote ocean area, the only major component of the shuttle system that will not be used again. The orbiter, meanwhile, continues its mission (it might last a week) and then returns to Earth, flying horizontally like an airplane in its last stages and landing on a runway near the launch site.

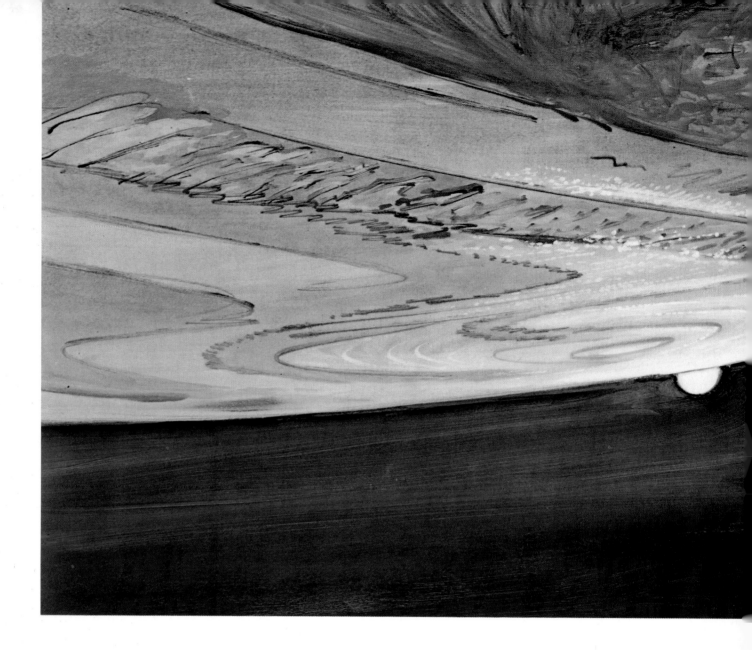

above: Shuttle orbiter in rendezvous with an Earth resource satellite. Not only will its own operation be much more economical than earlier space programs, but the shuttle system will also make vast savings by sending crews into space to repair malfunctions in orbiting satellites.

right: Cutaway of the advanced shuttle system in which both orbiter and main booster are manned. The booster, as the cutaway shows, is almost all fuel tank; like the orbiter, it would be flown back to Earth like an airplane.

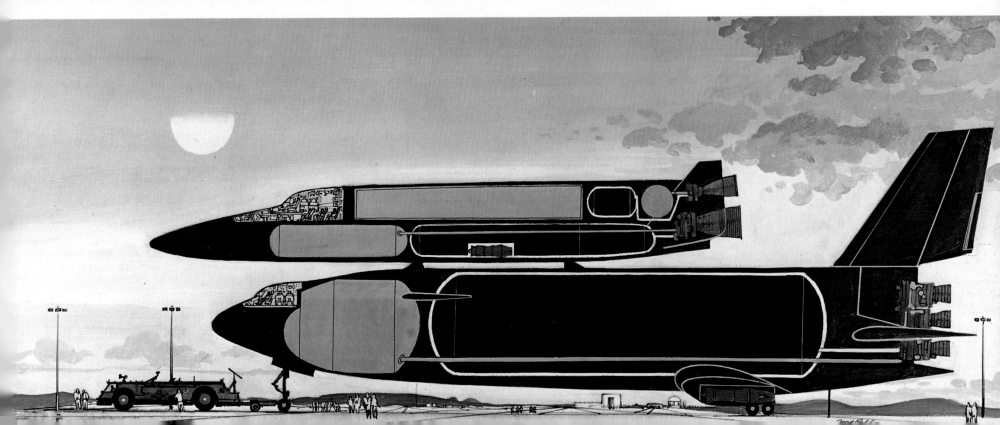

opposite, above: A shuttle in Earth orbit deploying its payload. The cargo has its own little thrusters for controlling attitude; these are operated by the astronaut at its right working at an external control panel. The tethers at the top of the painting extend from a space station that uses the shuttle system to replenish supplies.

opposite, below: Space shuttles at work. In the background a space station cluster extends its manipulating arms to assist a shuttlecraft in docking.

left: Detail of the shuttle orbiter in foreground of the painting at lower left: A payload specialist, visible through his round window in the nose of the craft, operates giant articulating arms that remove a satellite from the cargo bay to release it into orbit—a much more economical way to send an object into space than to launch it from an Earth base with its own expendable booster.

opposite: **The space bottle. A 1955 concept of Dr. Wernher von Braun for a work capsule—really a small spacecraft—that will allow one astronaut to work in a shirtsleeve environment controlling an array of tools. He can maneuver the space bottle in any attitude, anchor himself to his work station, and after his task is completed, can return to the space station, dock (headfirst) at the port, and enter through the station's airlock. Such a capsule could remove many of the problems astronauts have when confronted with unplanned extra-vehicular tasks such as the Skylab repair job.**

below: **Repairman in space. An astronaut tethered to a two-man spacecraft designed for this kind maintenance works on the antenna from an orbiting satellite.**

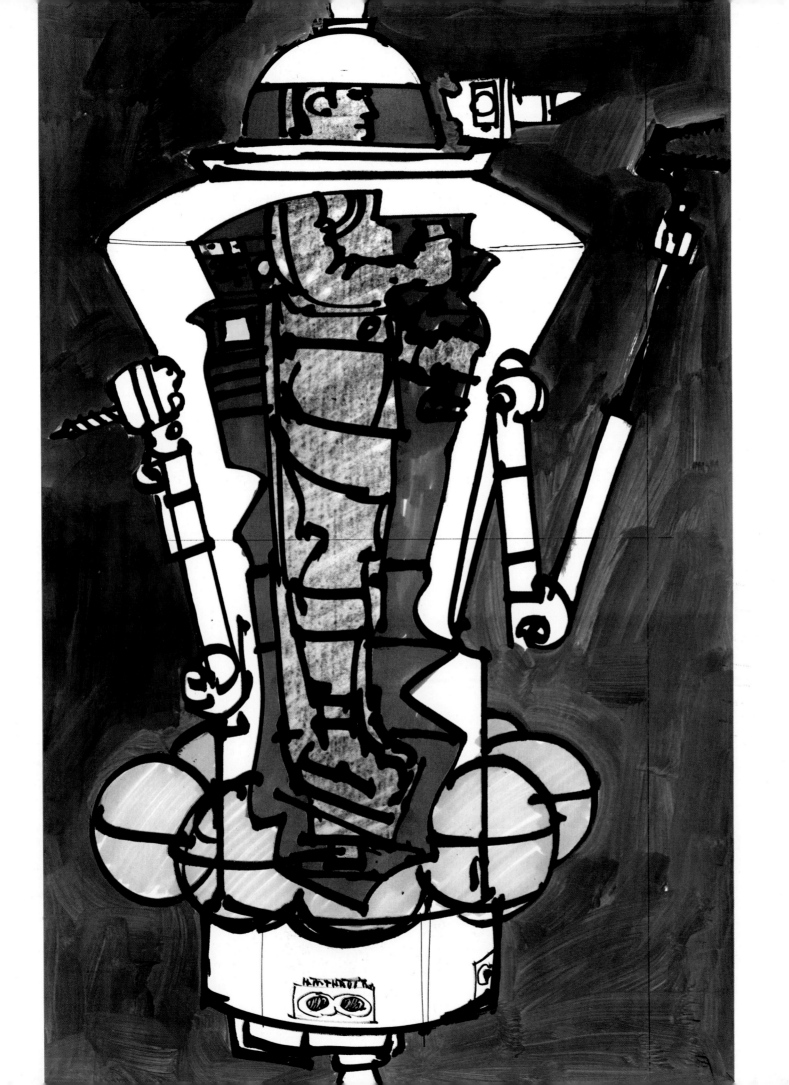

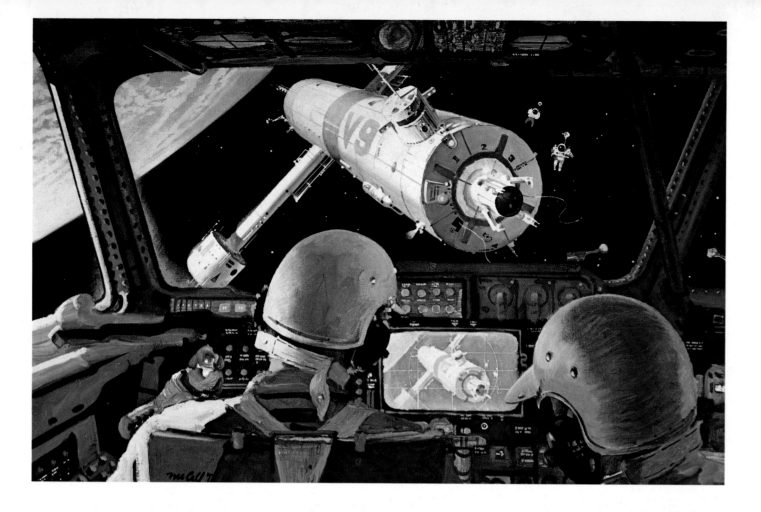

opposite, above: Inside the cockpit of a shuttlecraft, with the pilot and co-pilot preparing for docking with a space station.

opposite, below: The shuttlecraft docked with the station—in this case a top docking, but a nose docking is also possible. Two other shuttlecraft are seen, each of a slightly different configuration, since this scene looks forward to a time when shuttles, like aircraft today, will be specially designed according to their functions.

below: Astronauts on tethers working outside the space station; detail of painting opposite.

Small space station in orbit. Four solar arrays are deployed to provide power for the station, and below one of them is a maneuverable scientific instrument capsule. On top of the station is a high-gain antenna to communicate with Earth, and other communication and navigation antennas. There are four docking ports around the circumference at top and another four at bottom; a shuttlecraft is about to dock at the one seen below. The cutaway shows the living and working quarters of the crew; a central core connects the levels. In the lowest section (which has sleeping quarters at left) astronauts suited up for departure prepare to pass through the air lock and enter the shuttle, which will deposit the replacement crew and take the old crew back to Earth.

Transportation center at the turn of the next century. Under the great transparent dome of the terminal travelers watch one shuttlecraft take off, while another sits on its launching pad. A supersonic jet flashes across the sky, and in the right distance is a port facility with hovercraft. (Courtesy *Boy's Life*)

The Moon Colony

Up to now, 1973, all flights to the Moon have been made from Earth. Each covered a round trip of nearly half a million miles. Each group of astronauts remained on the Moon briefly and then returned, leaving nothing behind but some instruments, some debris, and some footprints.

This is an expensive way of conducting Lunar exploration, and not the best.

The lull scheduled for manned flights to the Moon after the Apollo 17 flight in December 1972 can be profitably used to establish a firmer base for such flights. Once that is done, the next group of Moon voyages can be made on a more massive scale. They will accomplish more with less difficulty and much less expense.

There are plans, for instance, to place space stations in orbit about the Earth. In a way they will be satellites, but satellites on a new and larger plan. Skylab is a beginning.

The satellites that have been fired into space periodically ever since Sputnik 1 in 1957 were placed into orbit in a single firing. Some of them were filled with sophisticated equipment and some were quite large—some large enough to hold men. None, however, was larger than the mass that could be hoisted into space by a single firing. And of those that held men, none was intended to remain in space for longer than a couple of weeks at most.

A space station, though, would be placed sufficiently high above the atmosphere so that it could remain in space indefinitely, and plans are under way for such stations. Men and women will be able then to remain in space for an extended period; space vessels will rise to the station carrying replacements and return with those who have completed their tour.

Reusable boosters will make the journey, each capable of living through a number of such tasks. This will, in itself, sharply cut down the expense of rocket launchings. (Imagine how expensive plane flights would be if a jet plane were used for one flight and then discarded.)

Each incoming vessel will carry material and equipment that could be added to the space station, which will thus become larger and more elaborate until it is far too large to have been launched in one piece by any rocket engines we are likely to build in the next few decades. It will become large enough eventually for the people working upon it to have a feeling of room. For the first time men in space will find themselves working under Earthly conditions (except for the fact that they will be experiencing zero gravity).

The fact that the space station is under zero gravity (because it is in free fall around the Earth) may be the most important aspect of all. At last we will be able to make extended experiments involving the effect of zero gravity on the body, because for the first time in space the body will not be immobilized, but capable of engaging in active exercise. (Bad effects of zero gravity experienced in small vessels where little movement is possible are not conclusive; immobilization is itself enough to produce deleterious effects.)

Animals can be subjected to tests that one might hesitate to apply to human beings. They can be mated and allowed to bear young to see how these would develop, both in the womb and in independent life after birth, if they had never experienced anything but zero gravity.

Such experiments might not only tell us things about the human body (about life in general, perhaps) that could be learned only with difficulty, if at all, in the presence of a perpetual gravitational field, but would also be a necessary preparation for the true exploration of the Solar system. All the bodies in the Solar system that man is likely to want to explore possess a gravity less than that of the Earth, and it would be wise to learn all we can about the effect of low gravity on man before we launch into a full-scale program of exploration.

Space stations can also be used to prepare vessels for flights to the Moon on a scale that would be impractical if Earth itself were the base. Such vessels could be brought to the space station piece by piece, not necessarily by manned flights. Unmanned conditions would be far less expensive, since inanimate pieces of metal can stand greater accelerations than fragile human beings can, and require far less in the way of safety precautions and life-support. Some of the pieces might even be fired into space out of a Vernian cannon in such a way that they could easily be trapped by the men on the space station.

Eventually the vessel and its fuel contents could be assembled just outside the space station itself. The assembled space vessel would be traveling somewhere between four and five miles per second (depending on the exact height of the space station) and would merely need to develop an additional two to three miles per second by rocket action in order to reach the Moon. A much larger payload could be carried for a given rocket thrust.

With one space station developed, spaceships traveling to the Moon could be equipped elaborately enough to begin establishing a second space station circling the Moon. They could even begin establishing a base on our satellite itself, one as permanent as the space station, and eventually much more commodious.

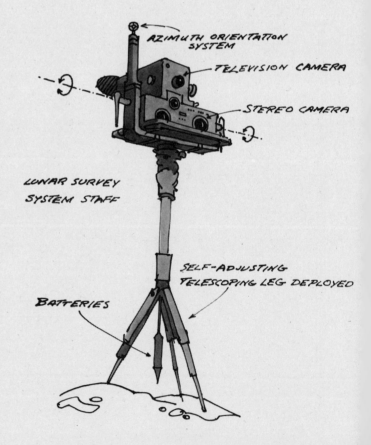

AZIMUTH ORIENTATION SYSTEM

TELEVISION CAMERA

STEREO CAMERA

LUNAR SURVEY SYSTEM STAFF

SELF-ADJUSTING TELESCOPING LEG DEPLOYED

BATTERIES

74

In short, the age of merely reaching the Moon in one-shot flights from Earth is over. With a space station in the sky, or perhaps more than one, the age of Lunar colonization will have begun.

And why not? Consider the analogous case of the discovery of the American continent by Columbus. On his first trip he did not merely come to America and return: he left a small manned fort on the northern shore of the island of Hispaniola.

To be sure, Hispaniola and the American continents seemed eminently suited to colonization, but there were difficulties—some of them much greater than those that seem to be barriers to the colonization of the Moon. To cross the Atlantic in the cockleshells then available meant a voyage of six or seven weeks, in any hour of which a storm might swamp the vessel. The distance to the Moon, on the other hand, can be covered in three days, with virtually no danger at all from the outside Universe. (The accidents that have occurred, whether fatal or non-fatal, have involved equipment failure.)

Psychologically, too, the advantages are all on the side of the Moon. The transatlantic voyagers of the 16th century traveled in complete isolation, with no communication with home at any time. The astronauts are in communication with Earth at virtually all times.

The very fact that the American continents were habitable meant that they were inhabited, and the natives on the spot presented a problem to the European would-be colonizers. When Columbus returned to Hispaniola on his second voyage a year later, he found his fort destroyed and the men gone—they had presumably been killed by the Indians. On the Moon, we are presented with what seems a virgin world. Whatever the dangers, hostile life-forms are not among them.

Then, too, the early voyagers to America could not know how large the American continents were or what conditions they would find. We know a great deal about the Moon. We know that it is not much smaller than the American continents. Its surface area is nearly 14,000,000 square miles, just a little less than that of North and South America combined, and over four times the area of the United States. What is more, we know nearly exactly what we will find on the Moon, and our technology is better able to cope with Lunar dangers than the European technology of Columbus's time was fit to cope with American dangers.

Yet with all the advantages Lunar colonization has over colonization of the American continents, there is no use pretending that the Moon is a paradise. It is rather the reverse. It possesses neither air nor water; and its period of rotation is so long that its day and night each last two weeks. With no water to absorb heat and with no air to distribute it, enough heat can accumulate on the sunlit portion

75

to create temperatures at the Lunar midday as high as the boiling point of water. The night temperature can drop to $-150°$ centigrade before dawn, far below the coldest Antarctic winter night. Without a protective blanket of air, moreover, the solid surface of the Moon is exposed to the harsh radiation of the Sun and a constant drizzle of tiny meteorites (few large enough, though, to do important damage).

This sounds horrible, but no one suggests that the Moon colony occupy the exposed surface. Place the Lunar station under the surface and at once many of the disadvantages of the Moon vanish. Temperature changes, as the Moon rotates, are restricted to the outer skin. The rocky crust of the Moon is an excellent insulator, and a few yards below the temperature remains constant and equable.

The same rocky roof that protects against temperature change will also protect the colonists from the radiation of the Sun and from the drizzle of meteoric matter. A large meteoric body, falling at just the right spot, would, of course, smash the colony at once, but such major collisions are very rare—and can happen on Earth, too.

But what about the absence of air and water? Would that not be disastrous below the crust as well as on the surface?

That depends on what we mean by air and water. There is no perceptible atmosphere clinging to the surface of the Moon, and no free-flowing or free-standing water in the form of rivers, lakes, or seas. That does not mean, however, that there is necessarily no water at all on the Moon. And if there should be hidden scraps of water, some of it could be split by an electric current into hydrogen and oxygen; the presence of water, then, would imply the manufacture of air.

Might there be bits of water here and there on the Moon?

Until men landed on the Moon, scientists could only reason indirectly. The three most common elements in the Universe at large are hydrogen, helium, and oxygen. Helium atoms form no combinations with each other or with other atoms and exist only singly. Hydrogen atoms readily combine in pairs, forming hydrogen molecules, and in the same way oxygen atoms form oxygen molecules. Hydrogen and oxygen have an even greater tendency to combine with each other, forming water molecules made up of two hydrogen atoms and one oxygen atom each.

For that reason, the most common compound (a substance with molecules containing more than one kind of atom) on any planet formed out of a representative mixture of the primordial material of the Universe is sure to be water. The only substances existing in greater amounts will be the elements (substances with molecules made up of one kind of atom only) hydrogen and helium.

This is the case on planets such as Jupiter, for instance.

Some elements and compounds are made up of atoms or molecules that cling to each other tightly through electromagnetic forces. They are solids even at fairly high temperatures. Such substances cling together even in the absence of gravitational fields.

Other elements or compounds are made up of atoms or molecules that do not cling together tightly. They are gases or easily evaporated liquids. Such gases and liquids are held to a planet primarily by gravitational forces.

Small planets, with small gravitational fields, are less able to hold on to these "volatiles" than are large ones. Thus, while Jupiter manages to hold on to a huge hydrogen and helium atmosphere, Earth has lost any helium and most of the hydrogen it may once have possessed. It retains enough hydrogen in combination with oxygen, however, to make up a large ocean, and it also retains an atmosphere of gases with molecules heavier than those of hydrogen and helium and therefore more easily held by a not-too-strong gravitational field.

The Moon, with a surface gravity only one-sixth that of Earth, did not manage to retain any atmosphere at all; nor did it retain enough water to allow the existence of oceans, or even ponds. Nevertheless, it seems reasonable to suppose that out of all the water in the primordial cloud out of which the planets formed, some might have been trapped under the Moon's surface. It seems possible that there might be veins of ice underground on the Moon as on Earth there are veins of coal.

Failing that, there is still the possibility that water molecules remain in loose combination with the molecules making up the rocky crust of the Moon. Such "water of crystallization," which exists on Earth, would be retained more firmly than free-flowing water would be.

Let us begin, then, by supposing that water in some form exists on the Moon, and consider how a Lunar colony might be established under such circumstances.

In the beginning, of course, the supply of water on the Moon will not be of importance. The colonists will bring all essential supplies from Earth—tools to dig out the first primitive cavern, air, water, energy sources, food.

But as long as the colonists are forced to depend on Earth for everything, they will exist at the end of an enormously long, and terribly fragile, umbilical cord. They will be forever in danger of extinction through failure of supplies or through the unwillingness of Earth to bear the burden.

There will be a tendency, then, to make any supplies from Earth last as long as possible. How about energy, for instance, since that is basic to everything else?

One obvious source of energy on the Moon is the Sun's radiation. It is the ultimate source of most of the forms of energy we use on Earth (nuclear energy is the most important exception), but on Earth it is difficult to use Solar radiation as a direct energy source. Sunlight is often interrupted by clouds, fog, mist, dust, and every twelve hours (on the average) by night. Even under ideal circumstances much of Earth's Solar radiation is absorbed by the atmosphere. Then, too, Solar radiation is a very dilute form of energy; a large sunlit area must be used to collect it in useful quantities, and in the great population centers where energy is most needed, space is most limited.

The situation is completely different on the Moon. The Lunar day is two weeks long, and all that time the Sun shines down with pitiless and unvarying intensity, unbroken by the slightest wisp of cloud, mist, or fog, undiluted even by clear atmosphere. Furthermore, there is virtually unlimited space on the Lunar surface, space needed for no other purpose (for no matter how populous the Lunar colony gets, it will always be underground, barring occasional laboratories and observatories), and undisturbed by anything but the occasional small meteorite strike.

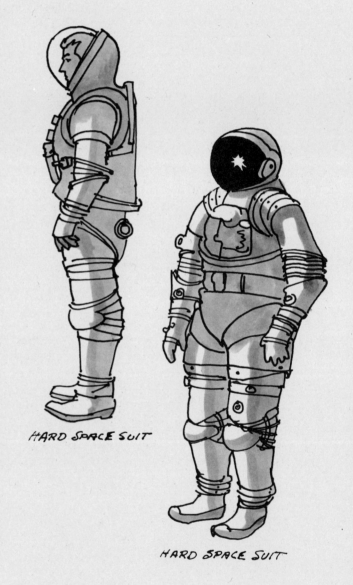

HARD SPACE SUIT

HARD SPACE SUIT

To be sure, there is a two-week night in which the Solar batteries, spread out over square miles of surface, will not be storing energy. In those hours, though, repairs and servicing can be carried through. It should be possible to lay down an area of batteries large enough so that enough energy is stored during the long day to keep the colony going during the long night. Eventually Solar energy stations will be established at several points on the Lunar surface, so that one or more of them will always be in sunlight.

Given an ample supply of energy, then, it will become possible to make everything else last longer. The air and water supply brought from Earth will not, after all, really be used up, in the sense that it will vanish after use. It will be merely converted to something else, which is "waste," because it cannot be used further—unless one goes to a little trouble.

The oxygen in the air is converted to carbon dioxide. Water is converted into such substances as perspiration and urine, or is simply dirtied by use in washing or in industrial processes. But it requires only energy to reverse the process. Carbon dioxide can be changed back into oxygen by appropriate energy-consuming chemical reactions. Good, usable water can be regained from liquid waste by energy-consuming distillation.

Of course to convert carbon dioxide back to oxygen chemically leaves us with carbon in one form or another (carbon that was once part of the bodies of the Lunar colonists), while to distill liquid wastes leaves us with various dry wastes. What does one do with these accumulated useless materials?

78

The problem will be largely solved if we consider food part of the cycle, for air and water in themselves are only half of it.

It is clear that food brought to the Moon to begin with will have to be as concentrated as possible and contain as little waste as possible. If the colonists bring merely processed food with them—packaged protein, plus some sugar and fat, minerals and vitamins—more will have to be shipped from Earth when the supply is used up.

In order to become independent of Earth, the Moon colonists will have to bring living things that can grow and serve as a continuing source of nourishment. Such living things are not going to be animals, since such animals would have to be fed also, and they would take more food than they would return when they themselves became food. The living food sources will have to be of a kind that will not itself require food in the ordinary sense.

That means the Lunar colony will have to live on plant life that will grow on inorganic substances. Furthermore, it will have to be plant life which is as nearly as possible completely edible. That would mean microorganisms of some sort or other.

We might visualize the existence on the Moon, then, of a cyclic process involving food, air, and water. The green cells of growing algae will reconvert carbon dioxide to oxygen; and out of that, plus the nitrogen, phosphorus, and minerals of human waste, the green cells will proliferate and form an expánding food supply. All that will really be used up will be the energy that will go into bathing the algae in light, since it will be this light energy the algae will use to reconstitute oxygen and food out of waste.

Naturally, the thought of an algae diet, like the thought of living in caverns, won't strike Earth-men as particularly desirable, but one can grow used to it. Men have endured worse. And as the Moon colony grows, there will be room for less concentrated food. It may even be that eventually rabbits or chickens can be raised on the Moon to supply the occasional meat meal.

Indeed, the Moon colony, which would be very primitive to begin with, may undergo numerous changes with advances in technology. The development of controlled fusion power might make it possible for the Lunar colonists to cease depending entirely on the Sun. With fusion power plants located underground, the colony might grow independent of the Lunar surface altogether.

Then, again, purely chemical methods may be developed for utilizing energy to combine water, carbon dioxide, and wastes to form oxygen and food. This may free the colonists of dependence on actual living systems other than themselves and increase the efficiency of the cycle.

Does this mean that, given a limited quantity of air, water, and food, plus unlimited energy, the Moon colony can go on forever, cycling and recycling and never running out of anything?

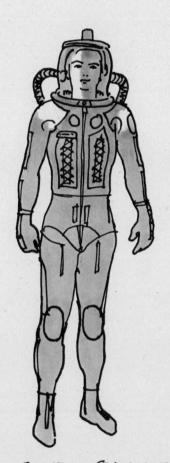

SKIN TIGHT SPACE SUIT

No. The cycling and recycling is never perfect, never ideal. At every stage there is some loss and this loss must be made up. Besides, it is only reasonable to expect the Moon colony to grow. What may begin as a tiny hole in the ground will develop into a more and more intricate system of caverns under the Moon's surface and end by becoming an underground metropolis. Indeed, we might even visualize it spreading out its tentacles until it is under every part of the Moon's surface.

But the larger the Moon colony grows, the more energy, food, air, water, and other materials it must pump into the cycle. And the larger and larger the cycle, the greater the loss through leakage that must be made up.

As far as energy is concerned, there is no problem. With Sun power and controlled fusion all reasonable requirements can be met. Food would be no problem, either, for given enough energy, carbon dioxide, water and minerals, cells will grow indefinitely. But can these substances be supplied in sufficient quantities?

When we speak of carbon dioxide, water, and minerals, we must consider as one source the Lunar crust itself. There is no question that there is an ample supply of many of the elements that will be needed. It is in those elements that form volatile compounds that the Moon is most seriously deficient—that means, in particular, the elements carbon and hydrogen. If carbonates, nitrates, phosphates, and water can be found on the Moon, then all is well.

The nitrates and phosphates (plus other materials needed in smaller quantities and surely present in the Lunar crust if the rest are) would serve as fertilizers; the carbonates would yield carbon dioxide on heating. As for the water, that would be valuable not only in itself but as a source of deuterium (or heavy hydrogen), the prime fuel in a fusion power station.

The supply need not be great, certainly far from the quantity that would be required by an Earthly population equal to that of the Moon colony. The Moon colony would be well engineered and well cycled, a population far removed from Earth's societies, which have been living in the midst of riches so long that they have never developed the habit of frugality in the use of resources.

But it may be that volatiles, and water in particular, are missing from the Moon. The rocks brought back by the Apollo astronauts are uniformly low in the more volatile elements and show no trace of water at all. One conclusion is that the Moon's outer crust was exposed at some time in the past to quite high temperatures over a prolonged period.

Why that should have happened is not certain. One of the more dramatic explanations is that the Moon was not formed in the vicinity of the Earth, but closer to the Sun, so that its crust is of the hard-baked type we would expect of the planet Mercury. (Mercury, the planet closest to the Sun, is at times only one-fourth the distance from that body that Earth is.)

Mercury's orbit is moderately elongated, but perhaps the Moon's was once more elongated still. At its closest approach to the Sun, it would have baked; at its farthest withdrawal it would have come fairly close to Earth's orbit. In some way, Earth would have managed to capture the little planet and make a satellite of it. If this is what happened, the Moon, though *now* constantly in the vicinity of our own water-rich world, would itself, with its lower gravity and the higher temperature of its Mercury-like past (both factors accelerating the loss of volatiles), be essentially free of water and remarkably low in carbon, nitrogen, and phosphorus.

What then? Would that mean that the Moon might be explored and studied, but never colonized?

Certainly it would make the colonization more difficult—but not impossible, if Earth were willing to make the sacrifice of supplying the water out of her own wealth.

It is not too great a sacrifice. If there is a shortage of water at all on Earth, it is of fresh water, and that is only localized. The supply of sea water is ample. It is more than ample, indeed, for there is the continuing possibility that the polar icecaps may melt and raise the sea level some two hundred feet, causing the gradual drowning of our continental lowlands. Getting rid of any sea water would be doing ourselves a favor.

We can imagine sea water shipped to the Moon—once the colony is a reasonably growing concern—perhaps with more nitrate and phosphate added, and some of the sodium chloride removed. Once on the Moon, the sea water would be distilled to yield fresh water and valuable solids.

Nor would a cargo of sea water have to be transported with the care and at the expense needed to carry human beings to the Moon. It could be fired off in the cheapest possible way and methods for trapping it on its approach to the Moon could be worked out.

There might be some reluctance to part with the water (suppose the icecaps *don't* melt, after all; it would be water permanently lost), but the amount needed by the Lunar colony would be surprisingly small. Only enough need be supplied to replace the unavoidable leakage in the cycling process, plus enough to allow for some growth. The minimum supply required will grow as the Moon colony expands, 81 and under these circumstances the colony's growth-rate will depend on Earth's generosity. If Earth refuses anything more than the minimum, then, on a waterless Moon, the colony will not grow.

This situation may not last forever. As techniques in space exploration improve, it may become possible to seek the necessary volatiles elsewhere. On the whole, the Universe is far richer in small atoms than in massive ones; and it is precisely the small atoms (hydrogen atoms are both the smallest and—by far—the most abundant) that the Moon needs. *It* may be short of them, but other bodies are not, especially bodies farther from the Sun, whose lower temperatures made the retention of the volatiles easier.

One rather dramatic possibility involves the comets. These are essentially small bodies, born far out in the empty vastnesses of space, and made up, as nearly as we can tell, largely or almost entirely of frozen material that at ordinary temperatures would be gaseous. When any of them, for any reason, enters the inner reaches of the Solar system and approaches the Sun, its substance evaporates and sometimes, mixed with rocky dust, forms a huge tail.

Comets, rich in hydrogen, carbon, and nitrogen, at times approach the Earth-Moon system, and we can imagine Moon colonists waiting for them. The colonists may land on them and mine them hastily for what they can get. They may even be able to divert them from their paths (comets are not large objects in terms of mass) and place them in orbit about the Moon. It would be rather like Eskimos beaching a whale and securing a nearly unlimited supply of food for the winter. When the Moon colonists beach a comet, they may find themselves independent of the Earth for decades.

Under the lunar crust.

Of course, we must ask ourselves what this is all about. Why should we go to all the trouble of establishing a Moon colony? Why should we subsidize the colony, supply it with its initial food, energy, and material of all kinds; perhaps continue to supply it with sea water for decades? What's in it for us that we should strain our resources in this way?

There is the matter of knowledge to be gained.

We have brought back many pounds of Lunar rocks which astronomers and geologists will be studying and considering for years to come. Because the Moon has not been subjected to the continuing action of wind, water, and life, its crust has been almost undisturbed except for meteoric bombardment. As a result, the Moon rocks are far older than any rocks we can study on Earth's surface and should tell us much of the early history of the Solar system, and of the Earth.

So far the material brought back from the Moon has offered astronomers no ready answers to their questions about Solar-system origins and early development. Rather disappointing. But then we have landed, so far, only on six places on the Moon and scooped up but six surface samplings. We are at the mercy of fortune in such a procedure. Who can tell what vital piece of the jigsaw we have missed by one mile?

With a Moon colony in existence, the Moon's surface could be explored thoroughly. The very existence of caverns would allow a study of the sub-surface layers. We can be sure that we would get answers to many puzzles.

The moon also offers an almost ideal platform for astronomical observations.

In the first place, it is airless. This means that nothing ever clouds the sky, neither mist, nor fog, nor cloud; not even the "clear" atmosphere itself, which absorbs 30 per cent of the starlight falling on Earth. Nor can temperature differences interfere with observations in the way that varying temperatures of layers of the Earth's atmosphere do by diffracting light beams so that there is usually an unavoidable fuzziness to the image seen by telescope.

Using even a small telescope on the Moon, one could see the surface of a distant object, such as a satellite of Jupiter, more clearly than we can see it from Earth. To be sure, a rocket-probe could tell us more than any ground-based telescope on either Earth or Moon, but rocket-probes can be sent out only periodically and their work is, at best, temporary. Remember, too, that there is a vast Universe which cannot be reached by rocket-probes, either now or in the near future. The realm of the stars and galaxies can be studied only at a distance.

In the last quarter century, moreover, astronomers have grown increasingly interested in radiations other than light, and almost all of these are absorbed by our atmosphere. Radio telescopes based on Earth's surface can, indeed, study the microwaves to which our atmosphere is transparent, but for anything else satellite observations are required.

On the Moon, the entire range of radiations and particles will be open to direct study, and there is every reason to suppose that once a complete and complex astronomical observatory exists on the Moon, a year's observation will tell us more about the Universe as a whole (and about the Earth as a whole, for Earth can be more easily studied from the Moon than from within its own atmosphere) than we have been able to discover in all the history of man.

What's more, if things continue in the way they have been going, the Moon observatory will become more useful with each passing year. As Earth's population increases and its technology intensifies, the atmosphere grows increasingly dusty, man-made lights grow increasingly bright, and man-made radio waves increasingly widespread. Year by year it becomes more difficult to study the light and radio of the Universe. The Moon's surface will offer an escape from this man-made interference, especially on the far side of the Moon, where Earth and all it produces will be blocked off by the Moon's 2000-mile thickness.

Other special properties of the Moon will also aid the astronomer. Since its surface gravity is only one-sixth that of the Earth, it would be easier to assemble a very large telescope there, and the mirrors and lenses of that telescope will be less distorted by gravitational pull. In the end, the really large instruments will be on the Moon, not on the Earth.

Again, the Moon turns more slowly than Earth does. On Earth, a star moves one degree across the sky (thanks to Earth's rotation) in four minutes; on the Moon, the same star will move by the same amount in 114 minutes. Telescopes must be geared to follow the apparent turning of the sky if they are to produce sharp photographs. It will be easier to arrange for a gearing thirty times less rapid, and objects of study will, of course, remain in the sky for that much longer interval for any given instrument.

But is the extension of knowledge enough? Is it proper to spend billions of dollars for something that might be viewed as simply serving to amuse a few scientists?

The answer could be that the extension of knowledge has, in the past, invariably worked to strengthen man's control of nature and has always offered man a greater chance to make use of natural law to improve his condition. When in the 17th century men bent over newly invented microscopes to study little bits of life too small to see with the unaided eye, it might have seemed they were amusing themselves with what any practical man would have considered trivia. Yet out of this, two centuries later, came the germ theory of disease and the conquest of the infectious plagues that had harried man for all his existence on Earth.

Still, a Moon colony has more to offer than abstract knowledge. The Moon has properties that would be highly useful to technology. For instance, Earth's atmosphere interferes with many manufacturing processes. Its effects must be countered, with great difficulty, or must be removed and a vacuum produced. Producing a vacuum is expensive, and the more nearly perfect (or "hard") it is, the more expensive. Leaks are a problem, for even tiny quantities of gas might supply molecules that will interfere with the process involved.

On the Moon's surface there is an ocean of vacuum harder than almost anything we can produce on Earth and, once the Moon colony is established, free for the taking. Any technological process that

84

Lunar colony concept.

requires a really clean surface, one without adhering gas molecules, can be conducted with the utmost simplicity. Thin films of metals can be layered onto bulk material very easily, so that the manufacture of microelectronic components could be carried on more efficiently. Welding and vacuum distillation could be done more easily on the Moon than on Earth. In fact, the mere existence of a permanent hard vacuum will result in man learning to do many things that have not been done on Earth simply because they are not possible to accomplish in an atmosphere, and vacuums on a really large scale are impractical.

Again, the temperature on the Moon can go down to the point less than half the distance of Earth temperatures from absolute zero. The Moon's surface temperature can be as low as 120° K. (120° above absolute zero, in centigrade degrees), whereas Earth's average temperature is about 300° K. To be sure, night does not last forever, even on the Moon, but there are places within the crater walls near the Moon's poles where Sunlight never penetrates and where the temperature may be permanently infra-frigid.

On Earth, we have in the last quarter century developed a flourishing "cryogenic" technology, a technology making use of instruments that work best at liquid helium temperatures (4°K. and less). At such temperatures, many substances are superconductive—that is, they can conduct electrical currents with no resistance at all—so that super-strong magnets are possible, and super-delicate (and tiny) electronic switches become possible. The computers of the future may require cryogenic temperatures to operate at maximum efficiency.

On appropriate places on the Moon, it should be considerably easier to reach and maintain cryogenic temperatures than anywhere on Earth, and research on superconductivity can be carried on more intensively. Eventually, the largest and best computer "on earth" may not be on Earth, but on the Moon.

In reverse, the day-side of the Moon attains higher temperatures and a more intense radiation level than is achieved on even the hottest parts of Earth's surface. What's more, without a protective atmosphere, the full range of radiation from the Sun reaches the Moon's surface, unaltered and undiminished. Experiments involving the use of such heat and radiation, experiments which would be impossible on the Earth, would be child's play on the Moon. And, of course, the Sun could be studied from the Moon in far greater detail than it could be from Earth.

Considering all these properties of the Moon's surface, it does not require very much imagination to see a Moon colony becoming a highly specialized factory for the production of instruments and components that cannot be produced as cheaply, as efficiently, or, in some cases, produced at all, on Earth. The Moon colony might become

a 21st-century Japan, so to speak, and pay its own way with a vengeance.

The mere existence of the Moon colony may help us in several other ways, less immediate and concrete than in serving as a super-technological base, but in the end even more important.

With the development of agriculture and civilization, man's capacity to affect the Earth has increased. When he existed only as a food-gatherer and hunter, he could forage, waste, pollute, reproduce to the limit of his capacity, and Earth absorbed it all (although early man may well have contributed to the extinction of some large mammals such as the mammoth). In a practical sense, Earth was infinite.

Agriculture, however, meant altering the face of the Earth with canals and crops. It deforested some areas and produced semi-desert conditions. It decimated wildlife and altered the pattern of the ecology. Population grew, cities expanded, and more and more of the Earth fell under direct human sway.

Even so, it seemed that Earth remained infinite, and capable of absorbing the worst we could do—until our present generation. Now our capacities, rising in a logarithmic sweep, have brought us to the point where we *can* destroy much or all of life on Earth, leaving the planet with a capacity for supporting only a sharply limited ecology, if any at all; and we are actually in the process of bringing this about.

If we are to avert such a doom, if we are to reverse the process of stripping Earth of its resources, fouling it with pollution, destroying species and wiping out the wilderness, we must alter the habits of two million years and recognize the fact that Earth is finite after all. This will not be easy, for to behave as though Earth is finite means the denial of liberties we have long taken for granted (the liberty to have children at will, for instance, or to throw something away after we have used it).

Presumably we will stumble along, doing the best we can, and trying desperately to preserve the Earth and ourselves. The Moon colony, however, will live, from the start, in a finite world, one much more sharply and harshly limited than Earth is even today. Every drop of water will be precious, every breath of air will be accounted for, every mouth to feed will be numbered.

If a Moon colony is to exist at all, it must live in an entirely engineered world, a tightly cycled world, a thoroughly calculated world. And it will teach us how such a world can be fashioned and made bearable. We will have to learn the lesson, for it is precisely such a world, on a much larger scale and with a much more complicated ecology, that we must construct ourselves if we are to avoid disaster.

86   The Moon colony, then, will be Earth's school in this respect.

City beneath the lunar crust.

In one way, the Moon's environment cannot be adjusted to resemble that of Earth. The caverns of the colony may be at Earth temperature, possess Earth atmosphere, be saturated with simulated Earth scenes behind mock windows. Nothing, however, can be done about the gravity. Very likely nothing can *ever* be done about the gravity.

The Moon's surface gravity will remain one-sixth that of the Earth, so we must wonder whether the human body can adapt to long-term exposure to low gravity. There is, at present, no way of knowing if it can, but we hope it can, and we know no reason (as yet) why it cannot.

If man can adapt to changes in gravity, it seems reasonable to assume that it would be easier to adapt from a higher to a lower gravity than the reverse. A 180-pound man could become accustomed to weighing 30 pounds, perhaps, but once he was accustomed to weighing 30 pounds, he might find it difficult indeed to get used to weighing 180 again.

We can easily envisage a course of exercise on the Moon that will serve to keep muscles and bones up to strength, even large centrifuges designed to give men practice at living and working under an effect similar to Earth gravity. Nevertheless, it seems likely that the longer a man remains on the Moon, the more reluctant he may be to return to Earth.

This would be particularly true of babies born on the Moon. The weaker stresses of gravity may lead to the development of thinner bones and weaker muscles, perfectly well designed for the needs of Moon-life, but incapable of handling the body's own weight on Earth. Besides, to people used to life in engineered caverns, life in the open under a frighteningly blue sky, with unfamiliar plant and animal life, with unaccustomed changes in temperature and weather, could be unbearable.

It may be, then, that the Moon colonists, and the Moon natives in particular, will be permanently separated from us in appearance and attitudes. The Moon natives will still be men and women in the full sense, but probably few Moon natives will be willing to undergo the long period of acclimation in centrifuges that would be required for a visit to Earth—though visits of Earth-men to the Moon would be far easier, and even common. We might therefore be faced, eventually, with the existence of two varieties of men and women: the high-gravity type of Earth and the low-gravity type of Moon; we might even call them hg and lg.

The existence of two varieties might be useful, since one might be able to do what the other could not. Consider, for instance, the exploration of the Solar system beyond the Earth-Moon system. It presents a complex of problems that the mere task of reaching the Moon does not. A space flight between Earth and Moon takes only three days, but a flight from the Earth-Moon system to any other sizable body in the Solar system would take anywhere from many months to many years.

It is questionable whether we could easily face a voyage of months to even the nearer planets. The difference between life on Earth and life in a constricted spaceship is enormous, and the adaptation may be more than an Earth-man can make. A Moon colonist, however, would live not on the outside of a planet, as we do, but on its engineered inside. The caverns of the Moon colony will be very like the interior of a large spaceship and the transfer into a smaller spaceship will be a far lesser cultural shock.

Also, the lower surface gravity of the Moon means that a rocket can more easily escape from it. And the virtual zero gravity on board a spaceship will offer fewer problems to an lg-Moon-native than to an hg-Earth-man.

In short, it may well be that for the intensive exploration and (perhaps) colonization of the Solar system, we will have to depend on a Moon colony; that it will be only after the establishment of that colony that space will really be open to mankind.

One final note, though.

In one respect, a Moon colony, or any colony or combination of colonies outside the Earth, cannot help us. No one of them, nor all of them together, can help us solve our population dilemma. If anyone thinks that the important reason for exploring space is to find outlets for our expanding population, let him think again.

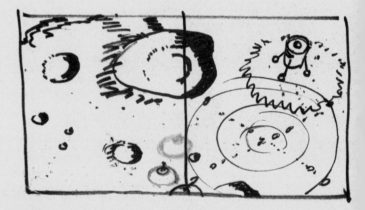

At the present time, man's numbers are increasing at the rate of seventy million a year. There is no way in which we can reasonably hope to place seventy million a year on the Moon, or anywhere else, within the next century at least.

And even if we could place any human being anywhere in space by merely snapping our fingers, and if every world in space were as warm and welcoming as Earth, the fact is that within five thousand years at our present rate of increase, the total mass of human flesh-and-blood will equal the mass of the known Universe.

In this respect, then, we must place our hopes nowhere but in ourselves. Space cannot help. Science cannot help. We *must,* of our own determination, and here on Earth, halt the population increase by balancing the birth and death rates. This can be done by increasing the death rate, but no one wants that. That leaves us with the necessity of decreasing the birth rate, and the existence of space will not remove that burden from our shoulders.

88   Remember that, above all.

3 Developing the Moon

preceding page: Rendezvous in lunar orbit resupplies a moon base—the glowing cross of lights at lower right. At the far end·of a nuclear ferry used to shuttle men and supplies from Earth orbit to lunar orbit, a stumpy "space tug" grasps cargo containers in its mechanical arms. The tug stacks the canisters on its top, then descends to the base to unload and return. Sunlight glints on a solar panel of a space station as astronauts tethered to it maneuver by individual rocket packs. All three crafts are in synchronous orbit above the Moon base.

right: Specialized vehicles will play essential roles in the exploration of the Moon that must precede extensive colonization. This scene—possible by the mid or late 1980's—shows an advanced lunar rover in the foreground, and in the background another type, one with a plexiglass bubble which would allow astronauts to work without cumbersome spacesuits. A space tug is parked at the right and another is just taking off.

this page, above: A plexiglass-covered rover on tractor-like treads delivers passengers at the portal of a hemispherical lunar housing unit. The accordion-action entry will attach itself to the bubble of the rover and allow the men to enter the shelter without changing into spacesuits.

this page, below: An astronaut operating an individual maneuvering unit flies miles above the lunar surface. The lights of a colony are visible in the crater at upper right.

opposite page: Three types of lunar vehicles. In the foreground a one-man flying vehicle with an antenna, a rocket system for forward thrust mounted aft, and a gimbaled rocket engine underneath. Right is a land rover with a solar panel for collecting energy from the Sun and a high-gain antenna for communication with the Earth. Flying in the distance is a lunar bus, a vehicle designed for long-distance travel and capable of carrying 10 to 15 passengers.

A city beneath the lunar crust. A vast area excavated by atomic power is supplied with an atmosphere that allows people to live much as they would on the surface of the Earth. At top center is a circular solar well reaching up to the surface of the Moon through which solar energy and sunlights are transmitted to the city below. The atomic power plant at upper right supplies the energy that is required by the city during the two-week lunar night. At left is a transportation facility; passengers will board the Earth ferry; once in Earth orbit they will transfer to an Earth shuttle that will take them to the surface.

Exploration continues: astronauts in the foreground on a surveying mission. On the lunar surface is a very advanced circular city; the spaceship above it is bringing passengers to land at the lunar city. (From *2001: A Space Odyssey,* © **1968 Metro-Goldwyn-Mayer Inc.)**

4 Mars

Where next, after the Moon? Either in manned probings while the Moon colony is being established, or in larger ventures after that colony is a working concern?

Presumably the next targets after the Moon are other objects that are close to Earth.

There are, indeed, some objects that are close to us, even closer to us than the Moon, that attain the ultimate in closeness by actually colliding with Earth as it sweeps through space. We are bombarded daily by billions of dust particles, some of which are large enough (pinhead-size will do) to develop enough friction to gleam as they pass overhead and produce what we call a "shooting star."

Almost all such particles are oxidized and vaporized in our protecting atmosphere, but some are large enough to begin with to survive the passage and, pitted and eroded and reduced in size, to reach the ground. These are meteorites, samples of outer-space material that man has been studying for nearly two centuries. (It was not until about 1800 that scientists could bring themselves to believe that solid objects fall from the sky.)

Some meteorites are large enough to cause devastation when they land; one big enough to destroy a city is not unthinkable. At least two have landed in the 20th century, fortunately both in uninhabited stretches of Siberia. During the billions of years that the Solar system has existed, most of the large fragments have been swept up (producing the craters on the Moon, for instance, and similar craters on Earth that have long since been eroded away by the action of wind, water, and life).

Nevertheless, a number of meteors yet exist in space, and one benefit of space exploration may be the establishment of a meteor watch, similar to the iceberg watch in the North Atlantic. From space stations, small bodies with orbits that carry them near Earth's orbit can be watched for, their motions calculated. If a collision seems even remotely possible, a hydrogen bomb (or some more advanced device) might be placed in the path of the body. Reduced to rubble, it would present Earth with no further danger; rather, a collision would produce merely a fine display of shooting stars.

From Earth's surface we can actually see some objects in orbits that can bring them within a few million miles of Earth, objects that are only splinters on the planetary scale, but quite capable of doing enormous damage on a human scale if they ever struck us, which, fortunately, is most unlikely.

Any small body that approaches us more closely than does any sizable body other than the Moon is called an "Earth-grazer." The first of these to be discovered was the asteroid Eros, detected in 1898 by the German astronomer Gustav Witt.

The orbit of Eros is such that when Eros and Earth are at the proper spots, they are separated by only 14 million miles, a distance only sixty times that separating Earth and Moon. Of course, the planets are hardly ever both in position for this closest approach, but every forty years or so there is a fairly close approach. In 1931, Eros passed us at a distance of only 17 million miles, and it will come close again in 1975. Eros is shaped rather like a brick, fourteen miles long and about four miles across. This is large enough to make a collision with Earth a major disaster, but there is virtually no chance of such a collision.

Other asteroids have been discovered that can approach Earth more closely even than Eros but these are, fortunately, smaller. The record approach for these came in 1937, when an asteroid, which was given the name of Hermes, was briefly sighted at a distance of only 485,000 miles from the Earth—just twice the distance of the Moon. There is some indication that at the proper point in its orbit Hermes would pass even more closely. However, its orbit could be calculated only with some uncertainty from that one observation, and it has never been seen again. It is probably less than a mile across.

It could be interesting to lie in wait for some of the "Earth-grazers" when they make a fairly close approach, get a close-up view, perhaps even land on them. There would be some interest in determining their shape and chemical structure. Instruments could be placed upon such asteroids, and they would then serve as mobile bases sending back information on those portions of space through which they pass. It would be desirable to use asteroids that pass through portions of space radically different from the regions immediately available to Earth and Moon.

The most suitable of the Earth-grazers for such a mission is Icarus, discovered in 1949 by the German-American astronomer Walter Baade. It is about a mile wide and can pass as close to Earth as 4 million miles. Its very elongated orbit carries it closer to the Sun than any planet comes; at its closest, it is only 17 million miles from the Sun. Any instruments placed on Icarus on one of its closest approaches to

Earth and picked up on the next approach might tell us a good deal about the space near the Sun that we could gain in no other way.

Comets can also pass close to the Earth, and their orbits are enormously elongated. Halley's Comet, which enters the inner reaches of the Solar system every seventy-six years, will retreat, at the far end of its orbit, to a distance of some 3000 million miles from the Sun. Instruments left on it at one close approach and picked up three-quarters of a century later may tell us something about conditions far from the Sun.

Undoubtedly, however, the asteroids and comets that pass near the Earth are not the most appealing targets. They are small, and they are, at best, infrequent visitors. What about the really large bodies of the Solar system? What about the other planets?

In one direction, that toward the Sun, there are two planets: Venus and Mercury. Their distance from Earth varies, of course. When their orbits bring both Venus and Earth in a direct line from the Sun, they are as close as they can get, and Venus is then 25 million miles from us—a hundred times farther than the Moon from us, but closer than any other large body (except the Moon) ever comes. As for Mercury, even at its closest it is twice as far from us as Venus is.

Venus is some 67 million miles from the Sun, compared to our 93 million miles, so that Venus receives about 2.3 times as much Solar radiation as we do. Mercury has an elongated orbit that places it anywhere from 26 to 43 million miles from the Sun, and it receives anywhere from 6 to 13 times as much Solar radiation as we do.

In general it is more difficult to go in the direction of increasing warmth than of increasing cold. For one thing, it is far easier to use energy to warm something up than to cool something down. For another, an approach toward the Sun and its warmth means exposure to the Sun's more dangerous radiation in greater intensities than are present here in the Earth-Moon system. Then, too, the closer we approach the Sun, the more expensive it becomes to maneuver in its intensifying gravitational field.

For all these reasons, mankind will be much more likely to explore space outward from the Sun rather than inward toward it.

Mercury, in particular, seems unrewarding. It is a small world, 3100 miles across, not much larger than the Moon. Furthermore, its structure is probably very much like the Moon's (and this is still more likely if the Moon's orbit originally carried it much closer to the Sun, in a Mercury-like approach). Like the Moon, Mercury is airless and waterless and, probably, cratered.

It rotates slowly so that every part of its surface gets 59 days of Sun and 59 days of night. It is conceivable that an exploring party might land on Mercury during the long night and burrow underground before Sunrise. Chances are, though, that at that distance from the Sun, the underground is no haven of moderate temperature but is still too hot for comfort.

Mercury's polar regions might be more promising, for there the Sun may always be near the horizon and some deep valleys may never to be exposed to its light. A base might conceivably be established there, and instruments set up outside the valleys in places where the light of the gigantic Sun may fall upon them.

The opportunity of observing the Sun at close quarters for an indefinitely extended period is the only obvious purpose of establishing such a base. Since some of the work might be done by means of the asteroid Icarus, even a Solar observatory might not be reason enough to dare the horrors of Mercury, and in the end mankind may satisfy itself with unmanned probes and the soft-landing of instruments upon the small world.

Venus may seem a far better target than Mercury. Venus is very close to us at times, receives considerably less radiation than Mercury does, and is protected from even that amount by a thick and permanent cloud cover which has generally, in the past, been assumed to be water. What's more, Venus is almost a twin of Earth in some respects, for its diameter is 7700 miles and its surface gravity is about five-sixths that of Earth.

Venus would be useless as an astronomic base, of course, since from its surface nothing could be seen through its clouds except for a general lightening to mark the difference between day and night. But because of its cloud cover, it once seemed possible that it might be a slightly warmer Earth, like an even wetter Africa, and it might turn out to be rich in life of fascinating (to Earth-men) differences and variety; and if it were not already occupied by intelligent beings, it might offer an extraordinary chance of colonizing a second world much like our own.

All such hopes have been blasted, however, in the last couple of decades. In 1956 it was discovered that Venus radiated radio waves characteristic of a warm body. It seemed that Venus might be hotter than anyone had thought. In 1962 a Venus-probe, Mariner 2, passed close to Venus and confirmed this. Since then other probes, plus a few soft-landed instruments dropped to Venus's surface, have placed the matter beyond doubt.

Venus has a thick atmosphere, some 90 times as dense as that of Earth, and 97 per cent of it is carbon dioxide. The carbon dioxide traps the energy of radiation from the Sun so that the surface temperature of Venus is 470°C., far above the melting point of lead. It rotates even more slowly than Mercury, but the devastating winds of its atmosphere distribute the heat so that all of Venus is at that temperature at all times, and nowhere, either on the surface or underground, is there any escape.

The surface of Venus is parched, and nowhere on it, even on its highest mountain peaks, can there be any trace of liquid water. What water exists on the planet is in its clouds.

There is therefore no likelihood of manned exploration of Venus. The exploration will be left to instruments, although one can visualize a manned spaceship taking a course that will dip it beneath the cloud cover of the planet for a quick look.

We turn our eyes, then, in the other, and much more hopeful, direction—away from the Sun.

In that direction the nearest planet is Mars, which circles the Sun at a distance of between 128 and 154 million miles and, at its closest, approaches to within 35 million miles of Earth.

It is a rather small world, 4200 miles in diameter—only half as wide as Earth, but twice as wide as the Moon. Its surface gravity is two-fifths that of Earth, just about midway between the gravity of the Earth and the Moon.

In other respects, also, it is midway between Earth and Moon. Unlike the Moon, Mars has an atmosphere, but an atmosphere that is far thinner than that of Earth, and one that contains no free oxygen. (Like Venus, Mars has an atmosphere that is mostly carbon dioxide.)

Unlike the Moon, Mars has water, or at least it has what look like icecaps around either pole. It does not have *much* water (if that is what the icecaps are made of), and there are no oceans on Mars—but it has some.

Mars is very much like Earth in its period of rotation, about 24.5 hours, and in the inclination of its axis from the perpendicular, about 24°. In both respects it almost duplicates our situation, so that the period of day and night and the positions and movements of the Sun in its sky would be very much like that of the Sun in our own sky. Of course the Sun is farther away, so that Mars' year and its seasons are nearly twice as long as Earth's. Furthermore, the more distant Sun delivers a little less than half as much radiation to Mars as to Earth, so that it is a considerably cooler planet.

In many ways, then, Mars is superior to the Moon as an object of human colonization. It is larger and has a higher gravity. It undoubtedly posseses carbon dioxide and water, and therefore other volatiles—not on the princely scale of an Earth, of course, but enough for the colony, so that it would be much easier than the Moon to make independent of Earth.

Since the atmosphere of Mars is not breathable, Mars colonists would have to live in enclosed volumes with a manufactured atmosphere. On the other hand, since there is an atmosphere that can protect against minor meteorites and against some of the Solar radiation (which is in any case only half the intensity of the radiation on the Moon), it is quite possible that the Mars colonists might choose to live in transparent domes on the Martian surface, rather than in caverns underground. This would make it possible to take advantage of the Earth-like day-and-night alternation and of the weakened Sunlight to grow the plants that would serve as food supply.

Energy sources might offer a more serious problem on Mars than on the Moon. Mars is the colder world, and whether the colony were on the surface or underground, more energy would be required for heating. The surface colony especially would require more energy, for at night the temperature routinely drops to Antarctic levels, while the deeper levels are not seriously affected. (Perhaps the living and working areas would be on the surface under the dome, the bedrooms well underground.) Then, too, it would be more difficult to use Solar radiation as a direct source of energy, since that radiation is less than half as intense on Mars as on the Moon and moreover the atmosphere does interfere. There are planet-wide sandstorms at times, for instance, that may block Solar radiation for weeks at a time.

If we assume, though, that by the time the question of colonization of Mars arises there will be controlled fusion power at hand, the problem vanishes. Energy from fusion can be obtained in any necessary quantity, and colonies on Mars, or on any even colder world that is otherwise suitable, can be maintained at comfortable temperatures without trouble. There is also the possibility, in the case of Mars, that the internal heat of the planet could be tapped, for it has volcanic regions.

From every angle, it appears that of all the worlds outside Earth itself, Mars is the one most suited for human occupation. The only disadvantage that cannot be removed is its distance. It takes three days to reach the Moon; six months to reach Mars.

On the other hand, Mars might prove to be so similar to Earth that it would be difficult to colonize after all. Might it not already be occupied by native intelligent life?

Speculation to this effect became serious in 1877, when Mars was making an unusually close approach to Earth and was nearly at its minimum distance of 35 million miles. An Italian astronomer, Giovanni V. Schiaparelli, observing it at that time reported the presence on Mars of straight dark markings which he called "*canali*," meaning "channels," because they looked as though they might conceivably indicate narrow bodies of water. Unfortunately the word was translated into English as "canals," which carries the implication of man-made waterways. At once speculation arose that Mars was the site of a very advanced civilization—a civilization of "canal-builders." The theory held that the planet was threatened with the loss of its water supply because Martian gravity was not strong enough to keep water vapor from very slowly leaking away into space; therefore these advanced inhabitants built mighty canals to carry water down from the icecaps in order to maintain agriculture and civilization.

This view was supported by the fact that there were dark areas on the generally ruddy background of the Martian surface that tended to wax and wane with the seasons. These were thought to represent plant cover which expanded in the summer of each hemisphere when its icecap melted, supplying water for the canals. The Martians were thought of as fighting a gallant and eventually doomed struggle against the encroaching desert.

In 1894, the American astronomer Percival Lowell established an observatory in Arizona; he studied Mars closely through that clear air and drew maps showing an intricate lacework of canals. Lowell was the staunchest supporter of the view of Mars as an abode of an advanced form of intelligent life.

Science-fiction writers were naturally stimulated to create an endless number of stories about Mars and its inhabitants. Opening the way in 1898 was the English writer H. G. Wells, in *The War of the Worlds*. Wells told of Martians who despaired of being able to maintain their world and launched an invasion of Earth. (All their ships landed on the small island of Great Britain, for some reason.) Their superior technology made it possible for them to overwhelm Earth-men, but they were defeated by physiology, for their bodies could not resist the onslaught of Earth's decay bacteria. It was the first story to deal with interplanetary warfare.

However, all the glamour of Mars as a dying planet inhabited by a super-intelligent race gradually faded. There were many astronomers who disputed the notion of canals altogether. The "canals" could just barely be made out, and only at those times when Earth's atmosphere was quiet enough to cause hardly any distortion of the telescope image. It was argued that they could easily be an optical illusion, for irregular spots too small and dim to be clearly visible would naturally be seen as straight lines by human beings straining to make them out.

Furthermore, the more closely Mars was observed throughout the 20th century, with more and more sophisticated instruments, the less likely did it seem that the planet could support any high-level civilization. The atmosphere proved thin and without free oxygen. The water supply seemed severely limited, and the very existence of canals seemed to grow steadily more dubious.

Nevertheless, it was not until 1965 that the matter was finally settled. In that year, the Mars-probe Mariner 4 (launched on November 28, 1964) sent back photographs of Mars taken from a distance of about six thousand miles. The photographs showed no signs of canals; nor have any of the more advanced probes sent to Mars since. There are no canals on Mars; no sign of great engineering achievements; no sign of any intelligent life.

That in itself is not conclusive. If photographs were taken of Earth from a similar distance, there would be no sign of intelligent life on Earth—or of any life. However, what the photographs *did* show made it seem unlikely that Martian conditions could support complex life-forms.

The photographs showed parts of the surface littered with craters, very much like those on the surface of the Moon. There was little evidence of erosion on the Mars craters, so it seemed reasonable to suppose that Mars had always been a world of very little air and water and that it was perhaps a world as dead as the Moon. Even the icecaps might not be water, but frozen carbon dioxide.

Even if this is so, and there is no advanced life on Mars, it is not a great blow. From the standpoint of human colonization, the absence of intelligent creatures, or even of large non-intelligent ones, removes one possible complication.

Furthermore, even with Martian conditions at their worst, there remains the possibility of simple forms of life—the equivalent of lichens or mosses, or, more likely yet, microorganisms. Such simple forms of life can survive even under the harshest conditions Mars would seem to offer. Indeed, simple forms of *Earth* life have survived under conditions mimicking those of Mars, and one would expect it to be much easier for native Martian life-forms, adapted to those conditions from the start, to exist under them.

Despite the depressing information sent back by Mariner 4, therefore, there is still talk of soft-landing instruments on Mars that would be capable of testing for the presence of chemical reactions associated with simple life-forms.

If life, however simple, did exist on Mars, it would offer a target of enormous value for a manned expedition, a target no less important, but offering far fewer problems, than would a planet with complex life-forms. Naturally, it is sad to be deprived of a chance of contacting a fellow-intelligence, but perhaps the enormous problems it would create might better be avoided at this stage in our explorations.

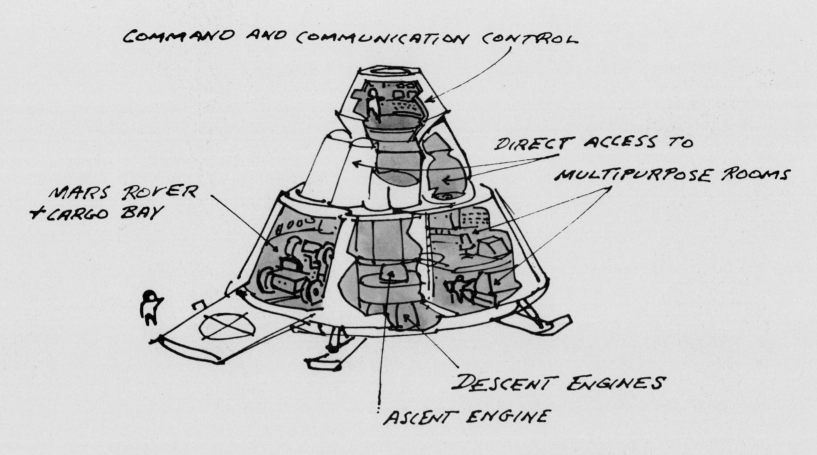

COMMAND AND COMMUNICATION CONTROL

DIRECT ACCESS TO MULTIPURPOSE ROOMS

MARS ROVER + CARGO BAY

DESCENT ENGINES

ASCENT ENGINE

MARS EXCURSION MODULE CONFIGURATION

If an unmanned probe confirms the presence of any kind of life on Mars, the pressures for a manned expedition would be enormous, for no instrument could investigate the nature of Martian life as well as a skilled biochemist could. The reason for this lies in the nature of life on Earth. Although it is tremendously varied, and although it exists in millions of species, all these forms of life, from hummingbird to redwood tree, from cockroach to man, from eel to mushroom, and from virus to whale, are essentially one and probably arose out of the same original blob of life in the ocean a number of billions of years ago.

All forms of Earth-life, without exception, are based on the large molecules of proteins and nucleic acids; all use essentially the same sort of reactions mediated by the same sort of enzymes. They are all variations on one theme.

If there is life on Mars, however simple, it may exist as variations on another theme, and this other theme would be clear enough even if only microorganisms existed. At one stroke we would double the kinds of life we know, and perhaps at once gain a more fundamental understanding of the nature of life.

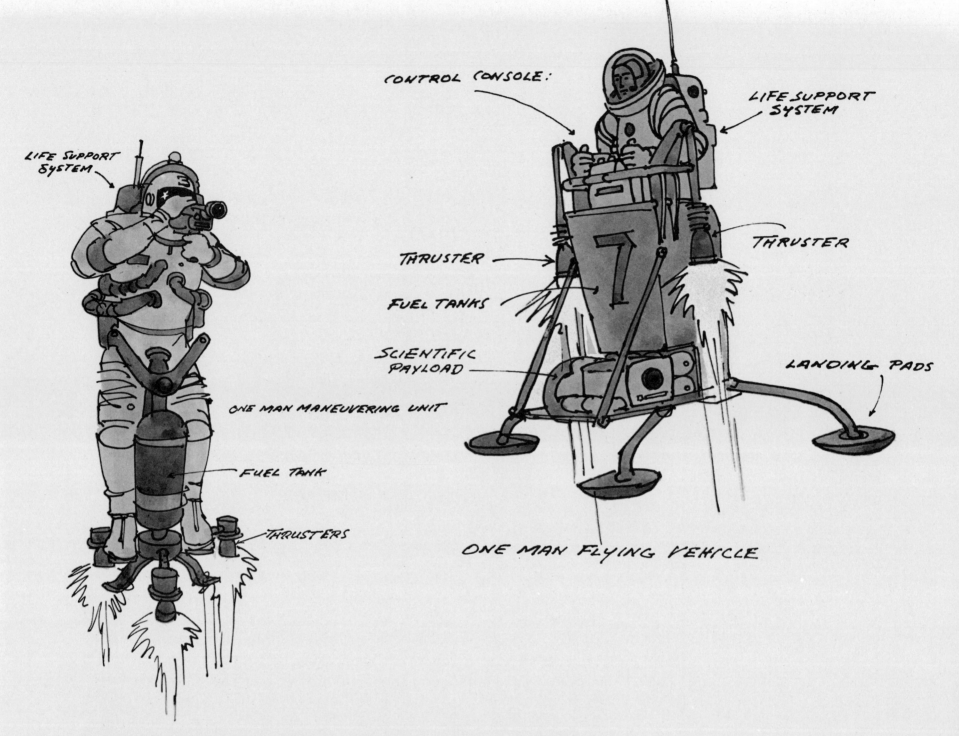

LIFE SUPPORT SYSTEM

FUEL TANK

THRUSTERS

ONE MAN MANEUVERING UNIT

CONTROL CONSOLE:

LIFE SUPPORT SYSTEM

THRUSTER

THRUSTER

FUEL TANKS

SCIENTIFIC PAYLOAD

LANDING PADS

ONE MAN FLYING VEHICLE

Even if life on Mars proves to be based on the same theme as Earth-life, there may be interesting differences in detail.

For instance, all protein molecules on Earth are built up of some twenty amino acids which (all but one) are capable of either a left-handed or right-handed orientation—one the mirror image of the other. Under all conditions not involving life, the two types, labeled D and L, are equally stable and exist in equal quantities.

In the proteins of life-forms on Earth, however, all the amino acids, with only the most insignificant and rare exceptions, are L. A uniform L-ness makes it possible for amino-acid chains to be built more neatly and stack more snugly, but a uniform D-ness would have done just as well.

Why L-ness and not D-ness, then? Is it a matter of chance, or will Martian life tell us that there is some basic asymmetry in nature that makes the L-form inevitable?

Or suppose Martian life turned out to be based on an Earth theme and to be identical even in detail. In that case, it might indicate that life as it exists on Earth offers the only biochemical basis of life on any world that even faintly resembles Earth.

Again, if Martian life were precisely like Earthly life, but were much simpler (as very likely it would be), it might be possible to discover something about the primitive basis of life that is obscure in the complex life-forms present on Earth. In that case, Mars would be a laboratory in which we could observe proto-life like that which once existed on Earth and even experiment with it . . . as on Earth we could do only if we owned a time machine.

Indeed, even if life did not exist on Mars, but if in its soil were half-formed molecules manifestly on their way to life-forms with biochemistry similar to that of Earthly life-forms, that might be helpful. It might indicate the nature of the path once taken on Earth, and we would perhaps learn much of the utmost importance to our biology and medicine.

And—to take it another step downward—even if Mars were completely free of life or even of chemicals on the way to life, there is still something certain to be of importance to us. The Martian atmosphere surrounds a world turning as ours does and exposed to temperature differences as ours is . . . but it is a thinner atmosphere, and one not complicated by varying quantities of water vapor, since there is no ocean on Mars. The pattern of its air circulation should therefore be simpler and easier to understand than that of Earth. What we learn of the circulation of the Martian atmosphere we might then apply to that of Earth, and come to understand our own weather better, learn to predict it more closely, and come closer to being able to control it.

(It would also be valuable, but harder, to study the circulation of the atmosphere of Venus. It is thicker than ours and different in composition, but Venus rotates so slowly that the effect of rotation can be ignored and thus the pattern of circulation is much simplified. Combining knowledge gained from Venus and Mars might tell us more about *Earth* than we could possibly learn from Earth.)

Interest in Mars increased suddenly as the 1970's began. On May 30, 1971, Mariner 9 was launched and sent out towards Mars. On November 14, 1971, it was placed in orbit about a thousand miles above the surface of Mars. This was not a matter of just flying by and catching what photographs it could; Mariner 9 was intended to

A rough visualization of a Mars colony concept.

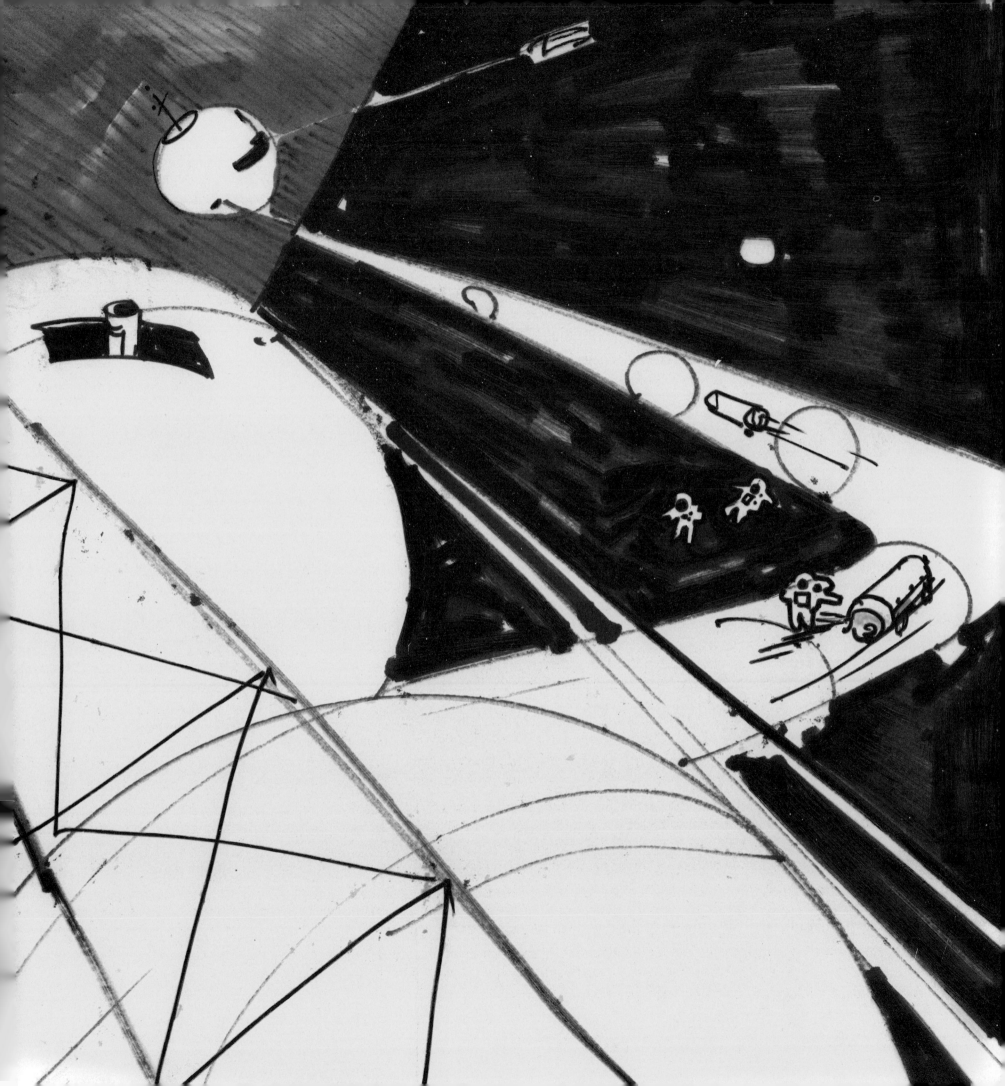

circle Mars indefinitely and take photographs for an extended period, eventually (if all went well) documenting the entire surface.

While Mariner 9 was on its way to Mars, a dust storm broke out on the planet and continued for months, obscuring the surface of the planet completely. Mariner 9 had to wait. By the end of December 1971 the dust storm subsided, and on January 2, 1972, Mariner 9 began to take its pictures. Eventually, the entire planet was indeed mapped, and it was quickly apparent that the limited sections photographed on earlier missions had not, after all, been representative of the planet as a whole. There were, it is true, large areas that were heavily cratered and seemed Moon-like in nature. But these were largely confined to one hemisphere of the planet; the other hemisphere was like nothing on the Moon, or on the Earth either.

The most startling feature seen was something which astronomers from Earth had made out as a small spot and called "Nix Olympica." It turned out to be a giant volcano, 310 miles wide at the base and therefore twice as wide as the largest volcano on Earth (the one that makes up the island of Hawaii). The crater at the top is 40 miles wide. Smaller volcanoes were noted in its vicinity.

Southeast of the volcanoes is a system of canyons which also dwarfs anything on Earth. They stretch across a distance equal to the full breadth of the United States; they are up to four times as deep as the Grand Canyon and up to six times as wide.

In one respect, Mariner 9 further dashed the hopes of earlier astronomers of life on Mars. The dark areas that expanded and contracted with the seasons were not vegetation after all. Some years before the Mariner 9 mission, the American astronomer Carl Sagan had advanced arguments for supposing the changes in the dark and light areas to be the result of dust of different kinds drifting this way and that with the winds. The Mariner 9 photographs, helped by the accident of the dust storm, showed that Sagan's views were quite correct. It was dust, not vegetation, astronomers had seen.

And yet the present thin-atmosphered, dry state of Mars may be temporary. There are signs that it cannot always have been so. Markings appear in the Mariner 9 photographs that look very like river beds. In fact, there seems hardly any way of accounting for the markings except to suppose that they represent what were once rivers, and even, perhaps, rivers that were flowing not very long ago, geologically speaking.

But if there was water not so long ago on Mars, where is it now? It cannot have leaked into space because the Martian gravity, working under cold Martian temperatures, is sufficient to hold any water it might now possess for extended periods.

The answer may lie in the nature of the Martian atmosphere, which is unique in one respect.

To understand this, consider the atmosphere of Earth and Venus. On Earth, some atmospheric components, such as water vapor, can precipitate out and become liquid water, or solid ice, if the temperature drops low enough. However, the oxygen and nitrogen that make up the bulk of the Earth's atmosphere remain gaseous at the lowest temperature Earth ever reaches. On Venus, the atmosphere is mostly carbon dioxide, which is easier to solidify than oxygen and nitrogen. Venus is hot, however, and never gets cold enough to solidify carbon dioxide. Earth and Venus both have permanent atmospheres, therefore.

On Mars, however, the atmosphere consists of carbon dioxide and water vapor, and at the lower range of Martian temperatures both can become solid. This means that virtually all of the Martian atmosphere could disappear and its solid components would be stored in the polar icecaps of that planet. Perhaps 99 per cent of the Martian atmosphere has frozen out in that manner. Or, to put it another way, if the Martian icecaps were vaporized, the Martian atmosphere might be as thick as that of Earth.

Close study of the Mariner 9 photographs of the icecap at the Martian south pole shows it to be terraced. There seem to be alternate periods of growth and recessions. It is possible, then, that for some reason Mars experiences long-drawn-out cycles, a kind of long winter and long summer in alternation. In the winter, the atmosphere is extremely thin and the planet is frozen and dry; in the summer, the atmosphere is thick, the planet is mild in temperature, and there are even some rivers.

In the long summer, there may be life on Mars; in the long winter, it may be quiescent. Right now, and for as long as Mars has been under telescopic observation, it has been in the long winter.

Suppose there is no sign of life on Mars whatever. A close study of the planet by Mars colonists might reveal the exact nature of the change that swings it from summer to winter and back again. It might be possible to start the cycle deliberately and create a kind of premature summer by keeping the cycle at that point, making Mars fairly comfortable for Earth-men. We might even imagine the planting of Earthly plant life that would, through photosynthesis, turn the carbon-dioxide atmosphere into oxygen and make it possible for men to live in the open as on Earth.

But what if there *is* life on Mars? Then there might be a question as to whether it is desirable to invade the planet and disrupt an ecology from which we might learn much. There might be an ethical question as to whether the planet does not belong to the life-forms that inhabit it, however simple.

It seems likely that in the end human curiosity and human selfishness would win out, and mankind would take over Mars. Some compromises might be reached. A period of time would be granted, let us say, for a thorough biological, chemical, and ecological study of the planet before intensive colonization proceeded—and then Mars' sister-worlds might be made use of.

Mars, it happens, has something that Venus and Mercury do not have . . . satellites. There are two Martian satellites, but neither is anything like our Moon; they are tiny objects, like asteroids in size.

The inner one, Phobos, circles Mars at a height of only 3700 miles above its surface, while the outer, Deimos, circles at a height of 12,500 miles.

While Mariner 9 was waiting for the end of the Martian dust storm, it took photographs of the two asteroids that made it possible to determine their sizes accurately for the first time. Both are irregular bodies: Phobos is 16 miles across its largest diameter, 12 miles across its smallest; the corresponding figures for Deimos are 8.5 and 7.5 miles.

Both satellites are pockmarked with craters, so that each looks remarkably like a potato in photographs, the craters representing the "eyes." Some of the craters are quite large considering the size of the satellites. The largest crater on Phobos is 3.3 miles across. The blow Phobos received when that crater was formed must have been just about as much as it could have sustained without breaking in two. In fact, it seems likely that the reason the satellites are so irregular in shape is that portions were broken off through collision with smaller bodies in earlier ages when space was more crowded with fragments than it is now. There is a cleft, or valley, a mile deep on Deimos.

These two satellites are almost like natural space stations from which Mars can be observed. Their usefulness is limited, however, because they revolve about Mars in the plane of the equator and give no useful views of the polar regions. To obtain an overall view, it would be better to build a space station considerably closer to the planet, one with a plane of revolution sharply tilted to Mars' equator so that all parts of the surface can be seen.

As for the satellites, perhaps they might be dedicated to a kind of Martian nature preserve. They might be hollowed out and a mini-environment constructed within in which the simple Martian life-forms might flourish. These could remain as a perpetual remnant of native life, to be studied, or merely to be allowed to live, while Mars itself became human.

112

4 The Path to Mars

Some time in the mid 1970's: The Viking spacecraft at rest on the surface of Mars. This unmanned mission scheduled for 1976 will obtain vital data and perhaps give some answer to centuries of speculation about the possibility of life on the red planet. The extended arm of the lander is scooping up a sample of Martian soil for analysis, while the television cameras in the yellow-topped poles swing around to scan the landscape. The data and pictures will be transmitted back to Earth by means of the mother craft (upper left).

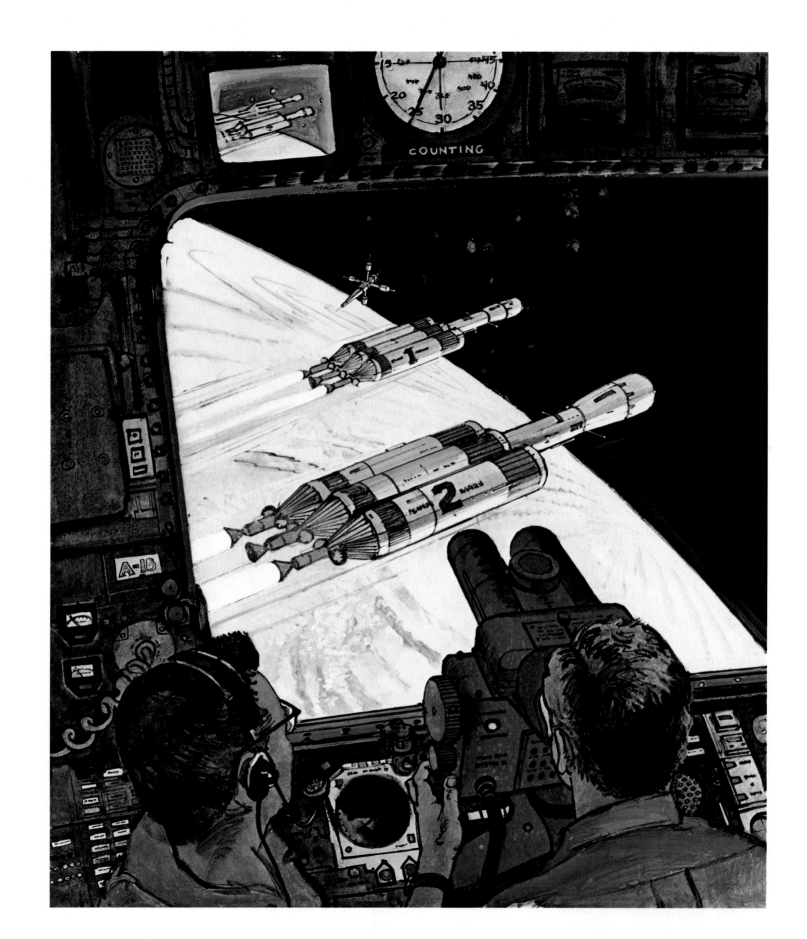

opposite: **The first manned mission to Mars, perhaps in the 1990's. Inside a space station in orbit two scientists observe the launch of the mission. The two spacecraft were assembled in Earth orbit and are now being simultaneously launched for the planet Mars. The outer boosters of each are firing, while the central one will remain inactive until the craft reaches the vicinity of the planet. There it might be used to adjust the orbit, and later it will be fired for the return journey. The nuclear-powered ships, each manned by a crew of six, will travel as a pair for safety's sake. Each is capable of accommodating the crew of the other in the event of a breakdown.**

below: **Scientists at mission control watch the first manned mission to Mars. The satellite Phobus looms large on the television screen, with the arc of the planet behind it. One of the two Mars craft is seen in orbit, its side boosters now jettisoned; the scene is presumably being televised by the second Mars craft.**

The first Mars lander has touched down, the cargo hatch has opened, and the Mars rover has rolled out. The second Mars craft, its descent engines flaming, eases its way down. Each of the landing craft has a crew of three, with three remaining behind with the mother ship–one of which is seen here in orbit.

opposite page: **A more advanced Mars base, perhaps early in the 21st century. A specialized Mars lander with an astronomical observatory on top is in the foreground; four other landers have opened their hatches and unloaded equipment—including some for the mining operation taking place at right. In the distance another Mars lander is about to touch down.**

this page, left: **A detail shows the ski-like foot of the huge lander and an astronaut riding a spacecycle. The large tank-like vehicle is a mobile laboratory.**

below: **Detail with another laboratory vehicle, this one with solar panels deployed. The Mars explorers wear spacesuits much like those worn by Apollo astronauts. Unlike our Moon, Mars has an atmosphere, although it is extremely tenuous and consists mostly of carbon monoxide.**

Spaceport, 21st century. Eventually space facilities of this type may serve as way stations for interplanetary travelers, while also acting as orbiting scientific bases, rescue areas, weather stations—even space factories for the manufacture of sophisticated goods that could be produced only in the vacuum and zero gravity of space. The illuminated portal is the launch deck, where spacecraft can discharge passengers and cargo and be launched back into space to continue their journeys to the Moon, the planets, or to return to Earth. This large station would be assembled in space from prefabricated geodesic elements.

5 Ceres and Jupiter

The colonization of Mars and its satellites will bring mankind to the limit of what we might consider Earth's immediate neighborhood.

The planets Mercury, Venus, Earth, and Mars, together with Earth's Moon and Mars' two small satellites, plus whatever smaller bodies move inside Mars' orbit for at least part of their own orbit, make up the "inner Solar system."

The inner Solar system is a small fraction of the whole. If the orbits of all the planets were mapped to scale, the space within the orbit of Mars would take only 1/6 of 1 per cent of the space containing the planetary orbits. The remaining 99 and 5/6 per cent is the "outer Solar system."

Those vast outer spaces contain five planets, compared to the four of the inner Solar system. Four of those outer planets are giants, with volumes anywhere from dozens to hundreds of times that of the Earth. The outer planets carry along with them twenty-nine known satellites, five of which are larger than the Moon, and a sixth that is nearly as large.

There is much to learn out there if distance will permit. The planet next beyond Mars is Jupiter, the largest and most fascinating of the planets, one that is so different from our own that given the opportunity, astronomers would be in agonies of curiosity to learn more about it. Jupiter's average distance from the Sun, however, is 484 million miles, more than three times the average distance of Mars from the Sun. From any point in the inner Solar system, the voyage to Jupiter is measured not in months but in years. It may well be that we will be forced to rely on unmanned probes (and the first of these, Pioneer 10, was fired in 1972) until such time as the Moon colonists, or the Mars colonists, or both, are ready to undertake the long voyages.

Nor does there seem much reason to hope that the duration of these voyages could be much reduced by technological advances to come. We may develop more efficient methods of propulsion and make the voyages cheaper, but the human body can endure only so much acceleration. If acceleration must be kept below a certain level and allowed to endure only a certain time, then the speeds attained must be limited, the time to attain them extended, and the voyage

Lift-off on Titan.

will be long. The Moon-men, who may be best fitted for long voyages, will be less able to withstand high acceleration than Earth-men (since the effects of high acceleration are like those of high gravity), and the voyages will be all the longer for that.

Yet there are targets short of Jupiter. Between the inner Solar system and the main regions of the outer, between the orbits of Mars and Jupiter, are thousands of small bodies—the asteroids. These are a little easier to reach than Jupiter.

The popular view of the asteroids is as a vast collection of small bodies. There is considerable truth to this. As of now, over 1600 asteroids have been sighted and have had their orbits plotted. Most of these are small bodies under ten miles in diameter. There are estimates that anywhere from 40,000 to 100,000 remain as yet undiscovered, mostly with diameters of a mile or less.

The fact, however, that so many are small should not obscure the fact that a few asteroids are fairly sizable. There are twenty-five asteroids known to have a diameter of over sixty miles and there may be many others. The largest known asteroid, Ceres, has a diameter of about 480 miles, nearly a quarter the diameter of the Moon. This may seem small, but it is not as small as it sounds. The surface area of Ceres is about the same as that of Alaska and California together, and that leaves plenty of room to explore, and, for that matter, in which to get lost.

The average distance of Ceres from the Sun is 257 million miles. It is only about 100 million miles from Mars at its closest approach, one-third the distance of Jupiter from Mars, and correspondingly easier to reach.

Is there any reason to grope outward for Ceres, however, aside from the mere achievement of doing so?

For one thing, Ceres might offer us a unique platform for astronomical studies, one which is, in some respects, better than anything we could find in the inner Solar system

Ceres, like the Moon, has the advantage of airlessness, and it has a surface gravity less than a quarter that of the Moon, or about 1/27 that of the Earth. A 190-pound man would weigh 7 pounds on Ceres. (That is another reason why the Moon colonists would be more suitable than Earth-men to explore the outer Solar system: they would be better suited to the low-gravity worlds they would encounter. No object in the Solar system on which human beings are ever likely to stand has a surface gravity over 4/10 that of Earth, and the vast majority have surface gravities equal to that of the Moon or less.)

One disadvantage of the Moon as a base for an observatory is the fact that for half the time the Sun is in the sky. This is valuable in that the Sun can be studied much more thoroughly from the airless Moon than from within Earth's soupy atmosphere, but the Sun's presence does cut down the time which can be devoted to the study of the stars and galaxies.

Ceres is 2.8 times farther from the Sun than the Earth and the Moon are. This means the Sun, as seen from Ceres, will have only one-third the diameter it possesses in the Lunar sky and will deliver only 1/8 the radiation of all kinds per unit area.

Indeed, all the airless bodies of the inner Solar system are near some larger body which represents some interference to the study of the Universe at large. Mercury is very near the Sun; the Moon is near both Earth and Sun; Phobos and Deimos are near both Mars and Sun. Ceres not only is farther from the Sun than any of these bodies, but it never comes closer than a hundred million miles to any object larger than itself. It is the most isolated body in the inner Solar system.

This isolation will not prevent an astronomical base on Ceres from offering a useful view of Jupiter every five years or so. Although one would expect the Ceres observatory to concentrate on stars and galaxies, its astronomers will at times be only one-third the distance from Jupiter that astronomers on the Moon are at its closest.

When Jupiter is at its closest to Ceres, it shines with a magnitude of -4.1, about the brightness of Venus at its brightest as seen from Earth. What is more, Jupiter's four large satellites would be seen as third and fourth magnitude stars, easily visible to the unaided eye when it is not blinded by Jupiter's much greater brilliance.

About the only characteristic, aside from distance, which makes Ceres a less satisfactory astronomical base than the Moon is the matter of rotation. Recent estimates show that Ceres' rotation takes a little over 9 hours. That means that objects in its sky will be moving from horizon to horizon about 2.7 times faster than they would in Earth's sky, and over 70 times faster over Ceres than in the sky of the Moon. Keeping a telescope in focus will be correspondingly more difficult on Ceres.

But can Ceres serve any other purpose than an astronomical base? Can one dream of a colony on Ceres?

Why not? In general, the colder a solid object has been in the course of its history, the more likely it is, all other things being equal, that it will have retained volatile material—including, most importantly, water.

The asteroid Vesta, for instance, the fourth asteroid to be discovered, has only about half the diameter of Ceres, yet appears brighter than

Ceres when viewed from Earth. Indeed, Vesta is the brightest of all the asteroids and is the only one that is bright enough, on occasion, to be made out with the unaided eye as a star-like point of light just on the border of the visible.

The logical conclusion is that Vesta reflects considerably more light than ordinary asteroids do, and this is most easily explained if we suppose it to be frost covered, at least in part. Ceres may not be as intensely frost covered as Vesta, but there is no reason to suppose that it was not as efficient at retaining water. It would seem, indeed, that the chances of finding adequate supplies of water on any of the more sizable asteroids is considerably better than the chance of finding it on the Moon.

Granted the presence of water in adequate quantities, Ceres offers an advantage to colonists that neither the Moon nor Mars does— or can. On the Moon, on Mars, and certainly on Earth, life is confined to the surface of the planet; but not on Ceres.

On Earth, for instance, it does not seem very likely that man will ever be able to establish living quarters very far beneath the Earth's crust (say, more than a mile below). If nothing else will stop him the temperature's steady rise with depth will.

The rise is probably less on Mars and the Moon, which are smaller worlds than Earth and therefore less compressed and heated at the center. Even so, the technological difficulties of probing more than a mile below the surface make such an endeavor impractical.

This means that almost the entire mass of large worlds is useless to man either as living quarters or as a source of raw material. It can serve as a possible energy source, though, and it is, of course, the source of a gravitational field.

But is all that mass absolutely needed for energy and for gravitation?

If it is possible to establish a permanent, ecologically independent colony on the Moon, that in itself implies that man can adapt himself to the absence of any significant gravitational field. To live on the Moon, man will have to go five-sixths of the way toward living without any significant gravitational field. If he can endure that, we might easily argue that he can endure a further drop to virtually zero gravity.

(This is by no means an iron-clad argument. Because man can live in an atmosphere somewhat thinner than that on Earth's surface does not mean that he can live in no atmosphere at all. On the other hand, it is easy to show that air—or at least oxygen—is essential to life; and no one has yet shown that gravitation is.)

If, then, men can live at near zero gravity, one reason to seek a large mass—which is otherwise useless for occupation or as a material resource—is eliminated.

From the standpoint of energy, the interior of a world like Ceres is hopeless. Under the pressure of a mere couple of hundred miles of rock, Ceres' interior is without the high temperatures that make Earth's interior (and even the Moon's, perhaps) a practical energy source. What is more, the Sun's radiation will be much less practical as an energy source on Ceres than on the Moon, since its radiation on Ceres is only 1/8 as intense as it is on the Moon.

Nevertheless (as with Mars), by the time mankind has reached the stage where manned expeditions can reach Ceres, it is almost certain that controlled fusion will have been developed as an energy source. And if the prospective Ceres colonists bring with them a fusion power plant, they will not require a large central mass for energy any more than for gravitation.

This means that there is no need at all for a body with an unreachable center. Ceres, in a sense, will be all surface. With a negligible gravity and with smaller pressures from above, there might be no limit to the depth to which funnels and caves could be driven. The entire body would be open to colonization, and if that is so, then Ceres, as a *habitable* planet, is far larger than it appears.

If we imagined Ceres split into parallel slices all the way through at one-hundred-foot intervals, the surface area of all those slices would be some twenty times as great as the entire surface area of Earth. Even if Ceres were not so thoroughly excavated, it is easy to see that, given an ample supply of energy and water, and supposing that the other elements could be obtained from the substance of the asteroid itself, the *available* mass of Ceres might be greater than that of Earth, Mars, and Moon combined and could support a population of millions.

Underground, of course, the gravitational pull is even less than on the surface, and we can imagine the colonists learning to take advantage of what traces of pull there are—learning to maneuver in flight by appropriate jumps and limb movements.

If Ceres is sucessfully colonized, if there is even the beginning of a successful colonization, then undoubtedly other asteroids will be in line for colonization next. The smaller asteroids may be more easily handled than Ceres, and one that is merely ten miles across may support a population of thousands.

The original colonists of Ceres may be Moon-men, who will have to adapt to still lower gravities, but once that is done, the asteroid-men will themselves colonize other asteroids. Each established asteroid colony will send out its frontiersmen, its Daniel Boones, in search of still newer worlds.

In the process, the colonies will work themselves farther outward from the Sun. The most distant asteroids of importance are the "Trojan

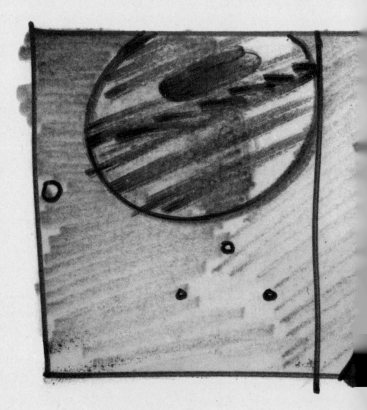

asteroids," so called because the first one discovered was named Achilles, and all the rest were also given names taken from the tale of Troy. They move in orbits virtually identical to that of Jupiter, one group located 60 degrees ahead of the planet, the other 60 degrees behind, and the two groups of Trojan asteroids together with Jupiter and the Sun form two adjoining equilateral triangles.

About a dozen asteroids are known in the two Trojan positions, and all must be fairly sizable or they could not be detected from Earth, which is never less than 300 million miles from them. There are probably others there which are smaller and therefore harder to see. When these are colonized by the asteroid-men, they would offer an even better base than Ceres for viewing the Universe, for they are more isolated. The intensity of Solar radiation on the Trojan asteroids is only 1/27 that on the Earth and Moon, and only 1/3 that on Ceres.

The Trojan asteroids would not offer a particularly good view of Jupiter, however, since they move in their orbit in step with Jupiter, remaining at a more-or-less constant distance of about 480 million miles. Earth itself is closer to Jupiter at times than any Trojan asteroid ever comes. On the other hand, an observatory on Achilles will, at twenty-year intervals, get a remarkably good view of Saturn, for it will then be roughly half as far from Saturn as we are at the closest.

The distance of the Trojan asteroids from the Sun, and the resultant cold, would not be a problem. Given fusion power, a colony could keep itself warm without trouble even if the Sun did not exist at all.

The asteroid colonists may end by becoming a third variety of human beings. There will be high-gravity man on Earth, low-gravity man on the Moon and Mars, zero-gravity man on the asteroids.

Zero-gravity men may become an incredibly numerous variety of humanity. It is not difficult to suppose that eventually a thousand asteroids will be colonized, with many millions occupying each of the largest ones, mere thousands on the small ones.

Undoubtedly not all the worlds will be alike in structure and composition, and we can imagine important trade among them. For example, comparatively water-rich asteroids (such as Vesta, for instance) might ship volatiles to those less well endowed, in return for, let us say, iron, in which some of the smaller asteroids may be rich.

The asteroids as a group, helping each other materially, may make themselves more habitable and capable of supporting a larger population than any one of them could alone. In the end, these small and

inconsiderable worlds, using trade to attain a fair representation of all the necessary elements and expending very little of them, thanks to efficient cycling, may support a population considerably larger than that of apparently larger worlds that offer only their relatively insignificant surface layers to human habitation. The zero-gravity people of the asteroids will considerably outnumber, eventually, the high-gravity people and the low-gravity people together.

But although the asteroid people can be an economic unit with respect to material resources (at least they most probably would be until their compositions are suitably adjusted), the colonies will nevertheless remain well insulated from each other by vast spaces as far as everyday life is concerned. Each one will surely develop its own ways, its own dialects, its own art, its own music, its own literature. There will be a cultural diversity among the asteroids that cannot possibly be matched on the surface of a larger world made into a tight unit by the advance of technology. The existence of a thoroughly colonized asteroid belt will therefore immeasurably enrich the life of mankind in every way.

And what of the Solar system beyond the asteroid belt?

The four giant planets, Jupiter, Saturn, Uranus, and Neptune, offer environments so extreme—atmospheres of enormous depth and pressures, gravitational fields of more-than-Earthly intensity—that it is unlikely that human beings of any kind will actually try to land upon them.

Even if we could imagine human beings remaining inside vessels designed, like bathyscaphes, to withstand enormous pressures, and imagine ourselves sailing through these atmospheres as we do through the abysses of the oceans of Earth, there would yet be extraordinary difficulties. The amount of energy required would be enormous, enough to brake safely through the gravitational field and enter the atmosphere at a velocity low enough to prevent frictional crisping, and then to lift out of the atmosphere and into open space against the gravitational pull—an energy requirement so enormous that men might well question the usefulness of the entire procedure.

This would be all the more true if low-gravity or zero-gravity men were to be the explorers, for they would be less suited to deal with a tremendous gravitational field and with the accelerations required for approaching and leaving. The alternative of sending unmanned probes into the atmospheres of the giant planets, or even ships manned by robots, would be so far preferable that one can easily imagine that no man will ever approach very near the giant planets, let alone enter their atmospheres. They, like Venus, will be forever off-limits.

128

Manned exploration of Titan, one of the 10 moons of Saturn.

There are, however, twenty-nine bodies of moderate size (all of them smaller than Mars) associated with those planets. There are twelve satellites of Jupiter, ten of Saturn, five of Uranus, two of Neptune. In addition, there are the independent planet of Pluto, possibly a number of as-yet-undiscovered satellites, and asteroids and comets that never enter the inner Solar system. There is no shortage of targets.

Venturing beyond the asteroid belt will be a task for both low-gravity men and zero-gravity men, and each will have different targets. This can best be demonstrated by considering the satellite system of Jupiter; of all the bodies beyond the asteroid belt, these are nearest, most easily reached, and most fascinating.

The zero-gravity men of the asteroids are, of course, nearest to Jupiter and its system. The targets appropriate for them would be small bodies with less than the surface gravity of even so small a body as the Moon, and bodies as far from Jupiter as possible, to avoid entanglements with the planet's gravitational influence.

As it happens these requirements are not hard to meet. No less than seven of Jupiter's satellites are in orbits that never bring them closer to Jupiter than six million miles and that take them out as

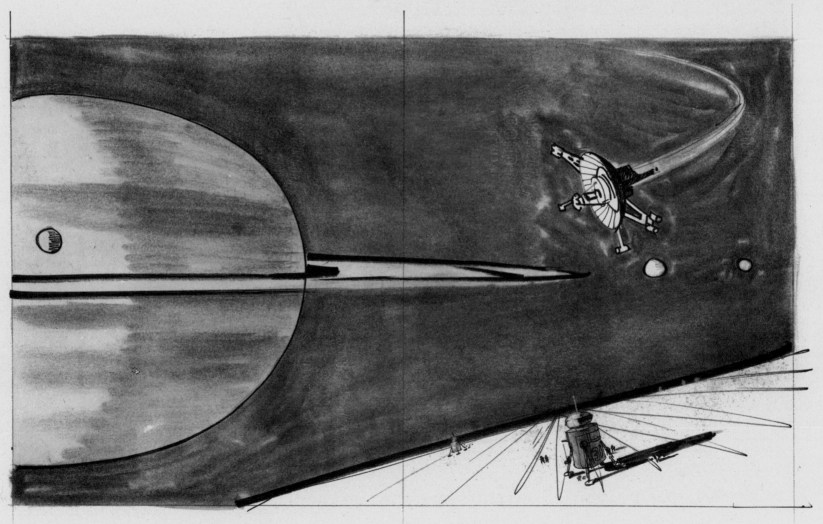

far from Jupiter (in one case) as twenty million miles. What is more, all seven are small bodies with diameters of from ten to seventy miles.

Ceres can approach as close to Jupiter as 200 million miles; other colonized asteroids will make even closer fly-bys. The Trojan asteroids may offer the best bases for a Jupiter-flight. Although always far from Jupiter (484 million miles, more or less), they are in the same orbit, at the same distance from the Sun. A rocket ship sent from a Trojan asteroid to Jupiter need not battle the Sun's gravity. It will merely be sent forward from an asteroid in the 60-degree-behind position, or backward from an asteroid in the 60-degree-ahead position. With scarcely any expenditure of energy, the ship will drift forward to Jupiter, or Jupiter will drift forward to it. It will take time, of course, but a zero-gravity man inside a spaceship would be subjected to conditions very little different from those at home and could take the time. And, because little energy need be expended, large payloads could be transported.

Colonists from the Trojan asteroids reaching the outer satellites of Jupiter will then find themselves on worlds very much like home . . . they will even be at the accustomed distance from the Sun.

This is no coincidence. Astronomers suspect that the outer satellites are captured asteroids.

The seven asteroidal satellites of Jupiter are usually known by Roman numerals representing the order in which they were discovered: Jupiter-VI to Jupiter-XII respectively. All may be colonized by the asteroid-men in the end, but perhaps the one they will first land on will be Jupiter-VIII, which is also known, unofficially, as Poseidon.

Of all the satellites, Jupiter-VIII has the most eccentric orbit. It is part of a group of four, all of which have average distances from Jupiter of from 13 to 15 million miles. The other three have orbits that carry them out as far as 17 million miles from Jupiter. Jupiter-VIII, however, recedes to a distance of 20 millon miles from Jupiter and reaches this point every two years. At that distance, Jupiter's gravitational influence offers no greater problem than our Moon's does, and it will be possible to land on Jupiter-VIII, and leave it, with little difficulty.

The colonization of Jupiter-VIII will offer no better view of the Universe than any of the colonized Trojan asteroids will, but it will give the astronomers among the zero-gravity people an excellent opportunity to study Jupiter at close range on a permanent basis. Even at 20 million miles, Jupiter-VIII base would be closer to Jupiter than we are even to Venus, while at the other end of its orbit, Jupiter-VIII swoops in to within 9 million miles of Jupiter.

At that close approach, which comes every other year, Jupiter will appear to the unaided eye about as large as our Moon appears to us; telescopic studies (even with instruments no better than our own) would be extraordinarily useful because of the lack of an atmosphere on Jupiter-VIII.

The close approach of Jupiter-VIII will offer the zero-gravity colonists a chance to make the jump to Jupiter's inner group of three asteroidal satellites with minimum expenditure of energy. These three, with an average distance from Jupiter of a little over 7 million miles, can attain distances of 9 to 10 million miles from Jupiter at the far end of their orbits. By choosing appropriate times, a spaceship from Jupiter-VIII could coast in orbit about Jupiter, without battling the gravitational pull, and reach the closer satellites. In fact, all seven asteroidal satellites could be colonized with little in the way of energy expenditure, once a base on Jupiter-VIII is established. From the innermost of the seven, Jupiter can be seen, at eight-month intervals, from a distance of less than 6 million miles.

Undoubtedly astronomers will want to get as close to Jupiter as they can. It is the giant planet of the Solar system, considerably more massive than all the other planets put together, a thousandth as massive as the Sun itself, perhaps more typical of planets in the Universe generally than the Earth is. The nature of its atmosphere and of its interior may tell us more concerning the Universe in general than anything in the Solar system other than the Sun itself.

Can we get closer to Jupiter, then, than the bases established on the asteroidal satellites?

Beckoning to men will be the large "Galilean satellites" (so called because Galileo first discovered them). These are, reading from the outermost inward toward Jupiter, Callisto, Ganymede, Europa, and Io. Even the farthest of these, Callisto, is only a little over a million miles from Jupiter, and the closest, Io, is only 240,000 miles from Jupiter—just about the distance of the Moon from the Earth.

Landing on these Galilean satellites and taking off again is bound to be more energy-consuming than maneuvering among the asteroidal satellites, for the closer to Jupiter, the more intense its gravitational field. It can probably be done, however, if we consider the degree of space-flight expertise men will have by the time they reach the satellite system of Jupiter.

A greater problem is the fact that the zero-gravity colonists of the asteroidal satellites will have difficulty landing on and exploring the Galilean satellites. These four satellites vary in size from a little smaller than the Moon to a size halfway between the Moon and Mars. Their

131

surface gravity is about that of the Moon, and the zero-gravity colonists may find it difficult to acclimatize themselves to that.

It may then be the low-gravity men of Mars or the Moon who will undertake the task of colonizing the Galilean satellites. To the Moon-men, particularly, the Galilean satellites may exert a powerful attraction, one that would justify the long flight outward. In size and gravitational pull the Galilean satellites will be like home to them. They are colder worlds than the Moon, of course, but that would not matter to an energy-rich society. The coldness is all to the good, in fact, for cold makes molecules sluggish, so that the Galilean satellites are certain to have retained far more in the way of volatiles than the Moon has. They have not been able to retain atmospheres, but they are certain to contain water. Indeed, recent studies make it seem that the satellites are frost covered.

Once the Galilean satellites are colonized, they may become water-exporters to the relatively dry asteroids—even, perhaps, to the Moon.

From the astronomic standpoint, the greatest value of the Galilean satellites will be the view they will afford of the giant planet, Jupiter. Each of the satellites, it is believed, rotates in exactly the same time it revolves about Jupiter, so that the same side always faces Jupiter. This means that from the Jupiter-side of each satellite, Jupiter will remain in a fixed position in the sky, rotating every ten hours so that all parts of it can be seen.

The Sun will cross the sky of each satellite, of course, and day and night will vary in length, from nearly twenty-four hours each on Io to over a week each on Callisto. Jupiter will show phases depending on its position with respect to the Sun. It will pass through the full stage during the time the Sun is below the horizon and it will then be a particularly magnificent sight.

Even from Callisto, Jupiter will have more than eight times the diameter that, to the unaided eye, our Moon has from Earth. From Io, Jupiter will have thirty-eight times the diameter of our Moon.

To astronomers, none of this may be enough. From Io, one would plainly see, even with the unaided eye, Jupiter's innermost satellite, Jupiter-V (Amalthea). Only seventy miles in diameter, it circles Jupiter a mere 70,000 miles above its surface, whipping about its orbit in twelve hours. Giant Jupiter would stretch a quarter of the way across the sky of Amalthea.

Nowhere but on Amalthea could one find so excellent a permanent base for the study of Jupiter. Surely efforts will be made to establish an astronomical observatory there. Not only would observers there be able to study the circulation of Jupiter's atmosphere with a detail possible nowhere else, but they could also study the properties of the space immediately neighboring Jupiter, that portion of space through which Amalthea is constantly moving.

From Amalthea, too, unmanned probes could be sent into Jupiter's atmosphere to send back information in detail on its composition and on the nature of the solid surface (if any) beneath the atmosphere.

And beyond?

Saturn's orbit lies 400 million miles beyond Jupiter. Farther on, 900 million miles farther on, is the orbit of Uranus; and another 1000 million miles beyond that is the orbit of Neptune. The gaps are enormous; each gap represents a voyage that would last years.

Although each of these planets is itself perhaps forbidden territory to human beings, who can no more land on them than on Jupiter, each carries planets with it on its lonely journey around the Sun. Each of these outer planets offers a tempting target for study and each set of satellites a tempting target for colonization.

Departure for Jupiter.

133

Saturn, particularly, is a fascinating world. It is a smaller Jupiter, in many ways, and one might suppose that with Jupiter under intensive study, there is little additional we could learn from Saturn to make the long outward voyage worthwhile. Saturn, however, has rings—something unique in the Solar system (and uniquely beautiful, too) which would certainly exert a powerful attraction on astronomers.

Saturn also has a family of ten satellites, some of which are no larger than large asteroids and would offer good fields of colonization for zero-gravity men. Most of Saturn's satellites revolve about the planet in the plane of its equator so that the rings are exactly at right angles to them. Since the rings are very thin, this means they are not visible at all from most of Saturn's satellites. (But the *shadow* of the rings cast on Saturn by the light of the Sun, which usually strikes the planet on a slant, can be seen.)

Saturn's outermost satellite, Phoebe, revolving in a plane tipped by 30 degrees to Saturn's equator, will offer the unaided eye a view of the rings of incomparable beauty. As Phoebe revolves, the rings will gradually change their slant, tilting first downward, then upward. Phoebe revolves around Saturn in eighteen months, so that every nine months the rings will be seen at maximum tilt (alternately downward and upward).

An accident in space.

Although Phoebe is 8 million miles from Saturn, that giant planet will seem at this distance about as large as the Moon appears to viewers on Earth, and the ring system will seem twice as wide. This sight will probably be the most beautiful the unaided eye could have from the surface of any world in the Solar system.

The innermost satellite of Saturn is Janus, just discovered in December 1967. It circles Saturn only 60,000 miles above its surface and only 12,000 miles beyond the outer edge of the rings. From Janus, it would be relatively simple to launch manned ships that would take up orbits at an angle to Saturn's equator from which the rings could be viewed broadside. Unmanned probes could be sent through the rings.

It appears (as far as we know now) that the rings may be made up of large fragments of solid matter that may be partly, or even mainly, ice. Some astronomers think that the innermost satellites of Saturn also may be made up largely of ice. It could therefore be that the day will come when the Saturnian system will be the water-provider of the colonized worlds; and that if there were ever a water shortage, Saturn could end it.

It is not very likely that men would actually undertake the long, long voyage just to carry a limited cargo of ice. That might not be necessary anyway. Fragments of ice from the rings might simply be

hurled into space toward the inner world and left to make the years-long trip on their own, following very eccentric orbits very much like man-made comets. Their orbits would be calculated and communicated to the inner world so that ships would be waiting to trap the ice and bring it down to the worlds that needed it.

Phoebe and Janus have diameters of a few hundred miles and are worlds of the order of the larger asteroids. This is true also of three more of Saturn's satellites, Mimas, Enceladus, and Hyperion. Three other satellites, Dione, Rhea, and Iapetus, have diameters in the neighborhood of 1000 miles, while Tethys has one of about 750 miles. Zero-gravity men may be able to handle them all.

By all odds, though, the most fascinating of Saturn's satellites is Titan, which circles Saturn at a distance of about 750,000 miles, completing its turn in about sixteen days. Its relation to Saturn is about the same as Callisto's to Jupiter, and it is about the same size as Callisto—about 3200 miles in diameter.

Titan's diameter is half again as large as the Moon's, and its surface area more than twice as large—about equal to Asia, Europe, and Africa combined.

What really makes Titan more interesting than the Moon, or than any of the Galilean satellites, however, is that its size, combined with its great distance from the Sun, has enabled it to retain an atmosphere. It is the only satellite in the Solar system known to possess an atmosphere.

Titan's atmosphere contains methane (with molecules made up of one carbon atom and four hydrogen atoms) and hydrogen. It is somewhat similar to the atmospheres of Jupiter and Saturn and to what may have been the atmosphere of the Earth when life was first forming. The atmosphere makes it possible for Titan to retain more heat than it otherwise could, so that the satellite is somewhat warmer than we might expect from its distance from the Sun, and it may be that life has formed upon it.

In fact, if we consider the other worlds of the Solar system, it seems that next to Mars, Titan is the most likely to offer us a sample of a living ecology to study.

Working outward in the Solar system from Saturn, there are the five satellites of Uranus, all with diameters of a thousand miles or less. Then comes Neptune, with two satellites. One is tiny Nereid, with a diameter of about 200 miles. The other, however, Triton, is a giant, another satellite of the size of the Galileans or of Titan. It, too, should have an atmosphere, but it is so far from us that astronomers have been unable to learn much about it.

We could suppose the low-gravity men and the zero-gravity men will divide the worlds of the outer Solar system between them. Low-gravity men from the Galilean satellites would work outward to Titan and Triton; zero-gravity men from the asteroidal satellites of Jupiter and from the Trojan asteroids would occupy the smaller satellites.

And beyond Neptune lies the lonely planet of Pluto, which has an eccentric orbit that is tipped quite a bit with respect to the orbits of the other planets. At its closest to the Sun, Pluto is only 2800 million miles from it, actually a little closer than Neptune ever approaches. At that time, however, because of the tipping of Pluto's orbit, Pluto is still nearly a billion miles from Neptune. (The ordinary diagram of the orbits of the planets makes it look as though Pluto's orbit crosses that of Neptune at one end, but that is a two-dimensional illusion.)

At the other end of its orbit, Pluto is 4500 million miles from the Sun.

Pluto is a lonely world indeed. It has no satellite as far as we know, and it never approaches within a billion miles of any other body, except perhaps for an occasional comet. The Sun itself is just a point of light—brighter than the full Moon seen from Earth by a good bit, but still only a point of light.

Perhaps Pluto, once it is reached, will offer the best platform of all for the study of the outside Universe. Far from the influence of the Sun and from the Solar wind that streams out of it, Pluto is at the edge of interstellar space. It rotates on its axis in 6.4 days, so that the objects in its sky move at only four times the rate of those in the sky of the slow-rotating Moon, and at only one-seventh the rate of those in our own sky.

In its grand sweep around the Sun, a sweep that takes 248 years, Pluto draws an ellipse that has an extreme width of 7300 million miles, in contrast to the 186 million miles of Earth's orbit. Given enough patience, astronomers on Pluto can measure the shift in apparent position of the stars ("parallax") with the change in Pluto's position far more easily than they can from any inner planet. This may give them a much more accurate notion of the scale of the Universe that will affect the general notions of its composition and development.

The size, mass, and gravitational pull of Pluto are almost exactly the same as those of Mars, as far as we know. If Mars colonists can manage the long haul to Pluto, once they have established themselves safely beneath Pluto's surface in well-heated and energized caverns, they may find themselves very much at home.

5 A New Environment

preceding page: A space platform, perhaps 100 years from now. This Earth orbiter (the arc of the planet is above), perhaps two or three miles wide, is the site of numerous scientific and commercial activities. At center an astronaut is being launched out of a port on a one-man mission. The huge spheres above are maneuverable cities with living quarters for hundreds of people. One of the many small spacecraft shown has just landed and is unloading passengers; they will enter the dome-covered areas where an Earth-like atmosphere is maintained. (Detail of this painting is on pages 14-15.)

this page: Departure from Europa. We are inside a space terminal on Europa, one of the moons of Jupiter; the arc of the planet dominates the sky, its red spot partly visible at upper left. A spacecraft is preparing for departure after having loaded passengers from the tower at right, which will be retracted before launch. A few local residents stand by to watch. In this very advanced community people are protected from the virtual vacuum by gear less cumbersome than that used earlier—here they wear small transparent spheres over the head and skin-tight suits under temperature-controlled outer garments. At right is the launch control tower, and orbiting above are two of Jupiter's twelve satellites.

Space wheel: Another space city concept, perhaps a century from now. In the foreground is a segment of the rim of the giant rotating wheel; in the background above is a similar city. The centrifugal force caused by rotation creates a simulated gravity at the rim, where most of the city's activities take place. As one moves toward the hub these forces decrease until zero gravity —weightless condition—prevails at the center. A rim bus suspended on a overhead rail (in right foreground) provides rapid transit on a circle route around the city.

Details of painting on preceding page:
this page: Cutaway of the spoke of the space wheel, showing control rooms, elevator core connecting the rim and hub, and astronauts at Level 4 preparing to exit through a lock for extra-vehicular activity.

opposite, above: A space hospital. Patients with certain illnesses might be brought from Earth to recover in the reduced gravity of this environment.

opposite, below: A small two-man sortie vehicle departs on some specialized mission.

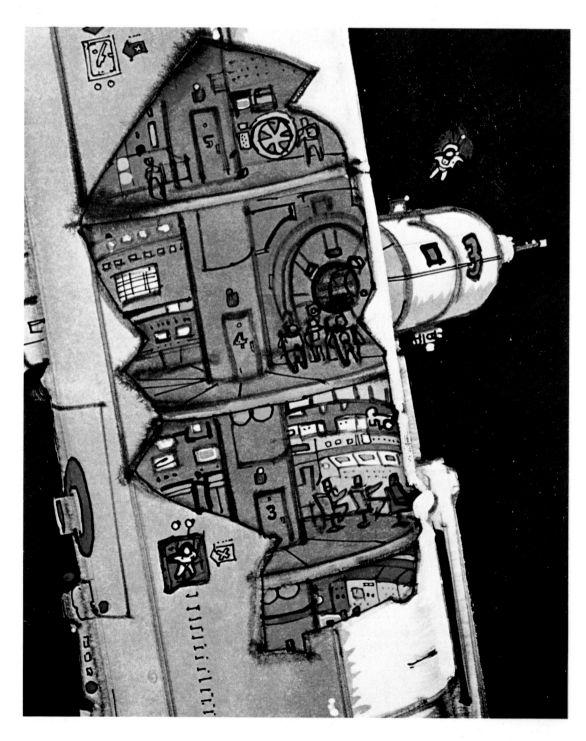

preceding page: **An accident in space. Astronauts at work retrieving a satellite that has perhaps been struck by a meteorite or damaged by an internal explosion. The view is from the inside of a recovery port of a very large space station in lunar orbit. One of the most valuable functions of such stations will be the recovering and repair of objects in space.**

this page, below: **Nuclear power and solar energy make the new environment workable. Here at left is a space city; on the right is a vast solar field that converts sunlight into microwaves and beams the energy to Earth, where it would be transformed into electricity, thus providing earth with an inexhaustible supply of pollution-free power. The one-man work modules at right are repairing meteoroid damage to the solar field.**

opposite: **Voyage to Jupiter, a scene from the film** *2001: A Space Odyssey.* **An astronaut (just visible through the window) maneuvers his one-man pod for some exterior task, while the huge spaceship "Discovery" drifts toward Jupiter. (***2001: A Space Odyssey,* ©**1968 Metro-Goldwyn-Mayer Inc.)**

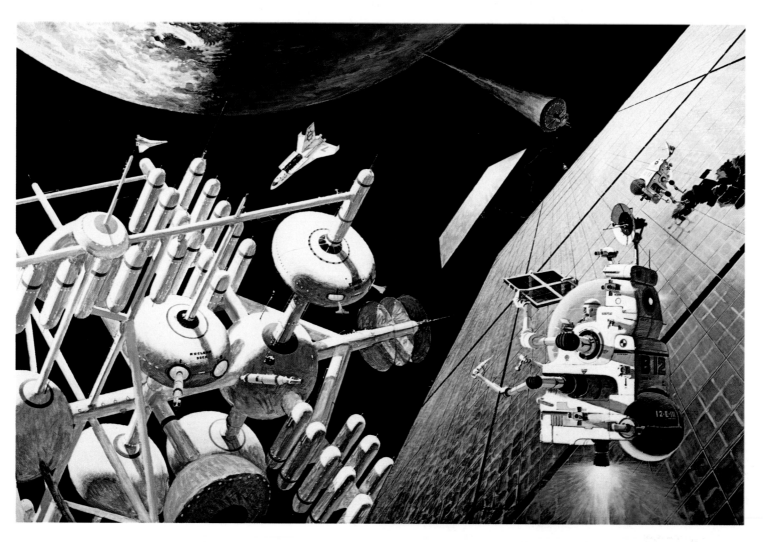

preceding page: A century from now a city on the Moon might look like this. Providing water on the arid Moon to sustain a colony this size would be a basic problem, but by this distant time we may suppose that this requirement will have been met. The huge hemispherical transportation modules would be self-sustaining and could house sizeable colonies of workers as well as moving them from site to site.

right: A highly developed city beneath the Moon's surface. The inhabitants dwell in an idyllic environment—perpetual daylight and perfect weather—and in the reduced gravity people move about with ease. Slow moving hovercraft buses provide transportation within the city center. The tall columns provide elevator service to the surface.

Detail of painting on preceding page: The artificially maintained atmosphere will permit the growth of plant life, and also permit lunar residents to wear light garments rather than elaborate space gear. Gradually the bodies of Moon-men will become adapted to a gravity one sixth that of Earth—and in fact natives of the Moon will probably have to undergo gravity-conditioning in centrifuges before making visits to Earth.

The Stars

Let us suppose, then, that by 2200 A.D. the Solar system is essentially colonized. Earth is still occupied by its high-gravity population. Low-gravity colonists have established flourishing worlds on the Moon, Mars, and the large satellites of the outer planets. Zero-gravity colonists occupy a thousand asteroids and a dozen or more of the small satellites of the outer planets. There are even scientific bases on Jupiter-V and on Pluto.

In a Solar system of this sort, the worlds will be, of course, independent and largely self-sufficient, since interaction among them (especially in the outer Solar system) will be both delayed and difficult.

Communication by electromagnetic waves is, to be sure, possible, particularly if it is assumed that technological advance by 2200 will make it possible to send modulated laser beams across interplanetary distances with little loss. Even so, it will take over an hour, at best, to send a message from Earth to a satellite of Jupiter and to receive an answer. To send and receive from Earth to Pluto, when Pluto is at the far end of its orbit, will take thirteen hours.

Granted the distance, the difficulty, and all else, there will not be complete isolation; there will be *some* sense of community; *some* outstretched threads of communication and fellow-feeling, however thin.

Then can the inhabitants of the Solar system, united (however precariously) and all contributing to scientific and technological advance, aspire to still further exploration of the Universe?

It might seem that now, finally, we must call a halt; that at Pluto we reach the end of the worlds to which mankind can fairly aspire. It may well seem that even attaining the most we can possibly expect of the technology of 2200 A.D., the Solar system must remain a permanent prison to mankind.

It is a matter of isolation; of fearful distances stretching outward that far transcend anything we encountered within the Solar system. The nearest star, Alpha Centauri (a three-star system, actually), is 25 trillion miles away. This is 5500 times as far from the Sun as Pluto is at its farthest. And that is the *nearest* star. A beam of light which may take 6.5 hours to reach from Earth to Pluto when Pluto is at its farthest, will take 4.3 *years* to reach Alpha Centauri, which is therefore said to be 4.3 light-years distant.

Of the better than one hundred billion stars in our Galaxy, only 39 stars or star-systems are closer than 17 light-years. The entire Galaxy stretches 100,000 light-years from end to end, with our Solar system 20,000 light-years from one of those ends.

Beyond our Galaxy are others. The nearest galaxies, two relatively small ones called the Magellanic Clouds, are about 150,000 light-years away from ours. The nearest galaxy which is like our own in size is the Andromeda galaxy, and that is some 2,300,000 light-years away. There are innumerable galaxies stretching still farther away, millions of them, probably billions, and some objects have been detected at distances of over a billion light-years.

These distances are, in themselves, quite beyond comprehension. We can try to approach them in another way, though, which may be clearer.

Scientists are quite certain that the speed of light is the fastest possible speed; that objects like ourselves cannot move at a speed faster than light. Since light travels at a speed of 186,282.4 miles per second, that might seem a rather generous limitation, one that need not concern us overmuch. Certainly as far as life on Earth is concerned, for all practical purposes, we might consider light as possessing infinite speed.

In the Solar system as a whole, however, the speed of light begins to be inconveniently slow. As pointed out at the start of this chapter, messages traveling at that speed take hours to make round trips between the worlds of the outer Solar system. This time lapse *cannot be reduced*.

Since the nearest star is 4.3 light-years distant, it would take any message, carried by laser beam, for instance, 4.3 years to go from ourselves to some planet of Alpha Centauri and 4.3 years for an answer to return. It would take nearly 9 years for light to make the round trip. That, again, is an absolute minimum, and that is for the *nearest* star.

To transport bulk material, including people, would naturally take much longer. If we traveled at the speed that would carry a ship along an elliptical orbit from Earth to Mars in six months, it would take perhaps 100,000 years to reach Alpha Centauri, the nearest star; 200 million years to reach the far end of the Galaxy; 6 trillion years to reach the Andromeda galaxy—and so on.

Under these circumstances, must we not resign ourselves to our Solar-system prison? Must we not submit to the isolation that these distances force upon us? And if we must, does it matter? Why would we want to venture out beyond the Solar system, anyway?

For one thing, of course, there is always the advance of knowledge; the chance of observing stars of different types from closer vantage points and in greater detail. We can see red giants, white dwarfs, neutron stars, double stars, pulsating stars, blue-white stars; and from them all we might learn a great deal more about the Universe than we might ever know otherwise.

For another, most, if not all, stars are accompanied by families of planets, and we might find endlessly repeated all the varieties of worlds—and more besides—that are present in our own Solar system.

Out there, there should be other Earths. According to some estimates the total number of Earth-like planets circling the various stars in our Galaxy alone (never mind the billions of other galaxies) is as many as 640 million. That surely should be of particular interest to Earth-men, when here in our own Solar system there is no world but Earth that will allow men to live on its outer surface, breathing a natural supply of air and surrounded by an ocean of water. It is only to planets of other stars that high-gravity people (as opposed to low-gravity and zero-gravity people) can spread without having to be subjected to an engineered environment and the necessity of adapting to a radically different gravitational force.

Then, too, on Earth-type planets there is a good chance that life will develop not only in the form of microorganisms, but also as vast, complex, and flourishing ecologies of large and varied species. There may even be a certain number of planets that have developed intelligent species. In fact, considering the number of Earth-type planets that must exist, the development of intelligent species here and there must be considered as certain.

If, then, we really want to take advantage of the biological and psychological information that can come of studying other planets like our own, as rich in life and mind as our own, it is out among the stars that we must go.

Of course, we might argue that if there is life out there and intelligence, there is no need to go anywhere. We need only wait and let them come to us. There are some enthusiasts and mystics who believe, indeed, that extra-terrestrial creatures have reached us at some time in the past. Others feel that we are being contacted even now by way of what are commonly called "flying saucers." None of the evidence for such contacts is convincing, but even if we are conservative and insist that no extra-terrestrial intelligence has impinged on us in the past or is doing so at the present, might they not in future?

However, if such a thing does happen in the future, it will mean that a race of intelligent beings has developed some practical method for traveling from star to star; and if such a method is possible, it is worth searching for. If they could do it, why couldn't we?

What are our options, then, and what routes might lead to successful interstellar travel?

To begin with, we might ask whether the speed of light is really an insurmountable limit to the speeds we can attain. If it is not, and if there were some way in which we could move at any speed we chose, then interstellar travel would merely depend on devising methods for achieving enormous velocities.

The first impulse is to say that the speed of light is a limit that is built into the structure of the Universe and that it cannot be surpassed. That, however, is a risky stand, for at many times in the history of mankind what had seemed flatly impossible under all conditions proved to be impossible only under certain conditions. When the conditions were altered, the impossible became possible after all.

In the case of the speed of light, that is a limit only for a particle possessing mass of any value between zero and the infinite. This may seem to include everything, but it is possible to manipulate the equations of relativity so as to allow them to deal with particles whose mass can be expressed by what mathematicians call an "imaginary number." This does not exist anywhere between zero and the infinite, but is outside the list of ordinary numbers altogether.

Particles with imaginary mass would not be bound by the rules that govern ordinary-mass particles and would be indeed able to move faster than the speed of light. In fact, they could *not* move at any speed *less* than light, so that the speed of light is a limit to them also, but in the other direction. There would be no upper limit at all.

These faster-than-light imaginary-mass particles are called "tachyons," from a Greek word meaning fast.

We might therefore imagine a spaceship and all its contents converted from ordinary particles into equivalent tachyons, moving to some very distant destination in some very short interval of time, and then being converted back into the ordinary particles of the original spaceship and its contents.

We might *imagine* it, but at the moment it requires extremely far-out imagining.

There is no evidence yet of the existence of tachyons, and they remain a mathematician's fantasy. And even if they existed, they would present staggering problems: how to convert all ordinary particles of a given mass to tachyons at the same incredibly small fraction of a time (for if some particles in the ordinary mass were converted more quickly than others, they would dash off to the other end of the universe without waiting); how to control the flight of the tachyons; how to perform the reverse conversion—the answer to each represents an achievement beyond even the dream of accomplishment.

So while we can leave the option of faster-than-light travel open, we must honestly admit that we cannot count on it. If we wait until the tachyon universe is brought to our service, we may wait forever.

While waiting, though, is there anything we can do *within* the speed-of-light limit?

Perhaps we can tackle the other side of the problem. If we are condemned to slower-than-light speeds, is there anything we can do about the distances? Are our objectives really as far away as they seem?

Yes, if our objectives are the stars. We must remember, however, that at great distances we can see *only* stars, because of their high temperatures and the vast energies they pour out. The nearest such object is indeed 25 trillion miles away and no sleight-of-hand will bring it any closer. But might there not be other subjects of comparatively low temperature, and therefore indetectable by conventional methods, that are closer?

Yes, there might. In fact, we can be quite certain that there are.

Comets that circle the Sun commonly have extremely elongated orbits. Halley's comet, the most famous of all, approaches within 55 million miles of the Sun at one end of its orbit (a little closer to the Sun than Venus is) but recedes to 3200 million miles at the other end, farther out than Neptune. The Grigg-Mellish comet recedes to 5000 million miles, farther out than Pluto.

These, however, are only among the farthest-sailing *short*-period comets (Halley's comet completing one turn of its orbit in 76 years, the Grigg-Mellish comet in 164 years). There are also comets—with periods in the millions of years, perhaps—which are one-time visitors to the Solar system as far as human history is concerned. These long-period comets must recede from the Sun to distances as much as a light-year or more.

Some astronomers have suggested that far out in space, well beyond Pluto's orbit and extending outward for one or two light-years, is a shell of asteroids formed from the outermost fringes of the vast cloud of dust and gas that condensed to form the Solar system. There might be a hundred billion such asteroids moving in slow and majestic circles about the Sun, each revolution taking 30 million years or so.

The gravitational influences of distant stars may sometimes so affect an asteroid of this sort as to cause it to fall in toward the Sun, even perhaps to enter the planetary portion of the Solar system. There,

157

planetary gravitation may trap it and keep it in the inner Solar system for indefinite periods of time. It then becomes a short-period comet.

It is not surprising that these distant comet-asteroids remain undetected from Earth despite their possibly vast numbers. On the average, they may be a mile or so in diameter, and all the bodies together may not be more massive than our Moon. They are so widely separated in the vastness of space that space remains virtually empty. They cannot be detected by the tiny bits of light they reflect, or by the tiny bits of light they block off, or by the tiny gravitational fields to which they give rise.

But they are there, and we might imagine zero-gravity men venturing out from Nereid, Neptune's small satellite, in search of such comet-asteroids, like new Columbuses facing a vaster deep. Perhaps if the Alpha Centauri system of stars is surrounded by a similar shell, the outer limits of the two shells may not be terribly far apart, and we can imagine asteroid after asteroid colonized, or at least used as a stepping stone, until there is a bridge between our Sun and Alpha Centauri, a bridge of slowly shifting worlds.

There are difficulties, of course. These distant comet-asteroids, formed far from any heat source, may consist almost entirely of volatiles. Indeed, when a comet enters the inner Solar system and approaches the Sun, the heat vaporizes its substance and surrounds it by a shell of dust and gas that is then hurled away from the Sun by the Solar wind, forming the tail that is the most distinctive part of the comet.

Concept for a space station, probably in Earth orbit.

It may prove impractical to attempt to establish a viable ecology on worlds too rich in volatiles. For one thing, warming such a world to the temperature human beings find comfortable might vaporize it. Then, too, even if a small number of the comet-asteroids were rich enough in non-volatiles and large enough to offer substantial homes to explorers, the distances between them would be enormous. The average distance between any two comet-asteroids is probably on the order of a billion miles; the distance between inhabitable ones, that is ones that are fairly non-volatile and fairly sizable, might well be more like ten billion miles.

It would be so difficult to inch toward the stars in this fashion, and so time-consuming, that though the explorations and colonizations might have value in their own right, they probably would not be looked upon as a way of getting to the stars. Rather than go by the stepping-stone method, it might seem better to figure out a method for covering the distance in a single jump.

Assuming, then, that the low-gravity or zero-gravity people will be the interstellar explorers of the future, we have now resolved (in some way) all the problems but one: who will build the space arks, and how?

The asteroid people suggest a possible solution. They are *already* occupying what are, in effect, space arks. We have assumed that it will be possible to engineer moderately sized asteroids (a few miles in diameter) into burrowed-out worlds with thousands of inhabitants and with independent cultures and ecologies. Larger worlds might contain populations of a million or more.

Suppose a group of asteroid-men fit some sort of rocket motor (or some particularly efficient descendant of such a thing, linked to their fusion reactors) to their asteroid-home and send it out of its orbit around the Sun?

What will these people lose?

After all, they will not have left home, for they would be taking home with them. Their entire society would be coming along—all their friends, their relatives, their entire culture. It would not matter how long the trip took, or where they went, or whether they ever got anywhere in particular. There would be no change in their immediate way of life.

To be sure, they would be leaving the Sun, but what of that? A dweller of the asteroids would not depend on the Sun for anything but perhaps the occasional chance of looking at it in the sky. He might miss that sight and idealize ''the Sun of home,'' but that would just be nostalgia and would have little, if any, practical influence.

The asteroid-men would also be leaving the other worlds with which they would have had contacts, cultural interchange, and, perhaps, the chance of help in an emergency; that *would* be a loss. However, the insulating space that separates the asteroids from each other and from other worlds would in any case have kept those contacts and that interchange rather small. To give it up might be painful, but it would not be a vital loss and might well be considered a reasonable price to pay for interstellar travel, especially since an asteroid that remained in orbit might find that others were leaving, so that the stay-at-home would end nearly as isolated as if it had decided to travel.

We might ask, though, why asteroids should pay any price at all for the privilege of making an interstellar trip. What's in it for them?

First, of course, there is the satisfaction of curiosity—the basic, itching desire to know. Why not see what the Universe looks like? What's out there, anyway?

Second, the desire for freedom. Why circle the Sun uselessly forever, when you can take your place as an independent portion of the Universe, bound to no star?

Third, the usefulness of knowledge. A trip of this sort is bound to add to the information possessed, and this new information could surely be used to add to the security and comfort of the asteroid.

Nor need such a journey be dull and uneventful. True, it may take hundreds or even thousands of years to reach a star, and generations may live without seeing one at close quarters, but does this mean there would be nothing at all to see? There is more to the Universe than stars, surely.

For one thing, it is very likely that any asteroid-ark on its way out of the Solar system will stop in the outskirts and pick up one of the comet-asteroids in the far reaches beyond Pluto. It will be hitched to the asteroid-ark and brought along for several very good reasons.

In the first place, it is a source of hydrogen, enough to keep the fusion reactor running indefinitely (and if one comet-asteroid is not enough, another can be picked up as well).

So to reach the stars, we have to accept both slow speeds and long distances. The combination means that any interstellar trip will take many years—indeed, decades and centuries, even to moderately nearby stars.

Is there any way we can fiddle with that time-lapse, even while accepting the limitations of speed and distance?

At first glance this may seem impossible, but duration, after all, is not only an objective measurement, but a subjective experience as well. We are not aware of the passage of time when sleeping or when under anesthesia; what, then, if we could sleep away most of an interstellar voyage?

Of course, our body ages during sleep, and sleeping away both the time and life itself is no answer. Aging, however, is an expression of the metabolic rate. That rate can be lowered as temperature is lowered, so we can dream of deep-freezing astronauts, who will then neither experience the passage of time nor undergo aging.

Astronauts, we can imagine, would be frozen after they have maneuvered the ship out of the immediate neighborhood of the Solar system and have set the course for some particular destination. They would then be thawed when they neared that destination.

Like the tachyon conversion, we have no way of knowing if this freezing scheme can ever be made to work. Freezing something as complicated as the human body without killing it, keeping it frozen for long years, then bringing it back to full life with no bad effects (even on the complicated, delicate, and all-important brain) is an enormous task. We have no reason for being confident that it can ever be done.

Fantasy of a gravity-free city of a distant future.

160

Then is there any way of fiddling with time without fiddling in a major way with the human body?

Yes, as a matter of fact. Physicists are quite certain that any object moving with respect to the Universe in general experiences a time-rate slower than does an object at rest. At ordinary velocities this slowing of time is infinitesimal and of no practical importance. At velocities in excess of one hundred thousand miles per second, the slowing

of the time-rate is easily measurable; and at velocities close to the speed of light, the time-rate slows down to the merest crawl. It would be zero *at* the speed of light.

Imagine, then, a spaceship heading out from the Solar system at a steady acceleration. Eventually speeds very close to that of light will be attained—at the cost, naturally, of enormous energy expenditures to make that long-continued acceleration possible. Once that

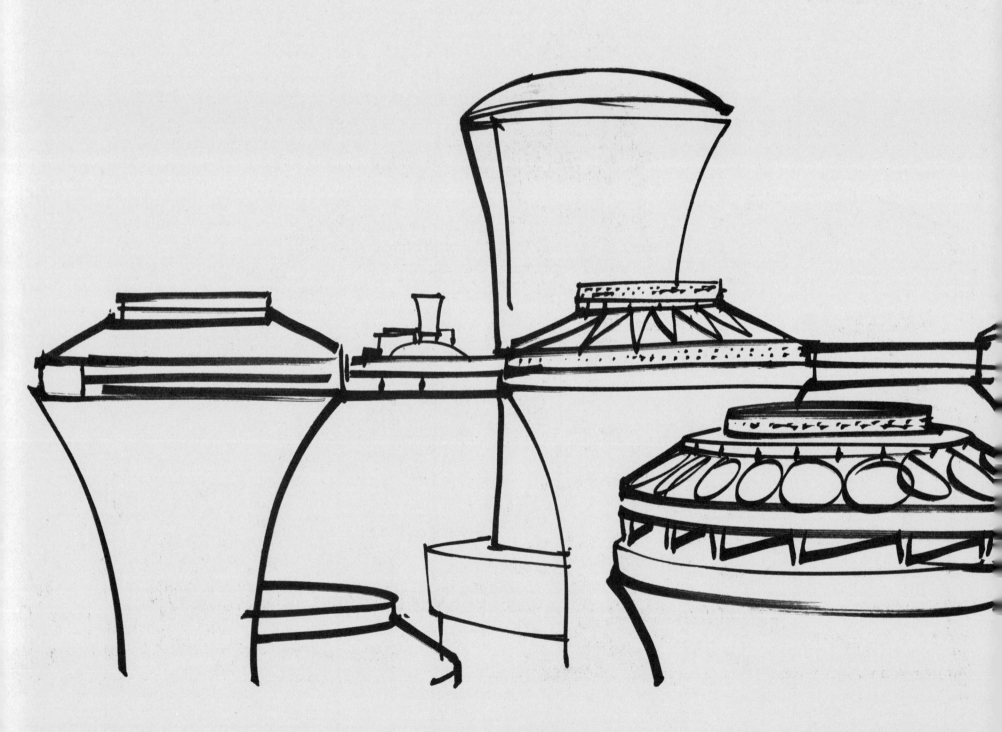

near-light-speed is attained, the ship can be allowed to coast. The time-rate will then be close to zero for the astronauts, so that it will seem to them that they are moving at almost infinite speed, and that any star they aim at (*any* star, no matter how distant) will be reached within a short time.

Of course as their destination approaches, they must begin to decelerate, a process that will take as long and be as energy-consuming as the original acceleration. Finally, when within planetary distances of their destination, they will be moving at ordinary velocities. The time-rate will then be normal for them, and they will be able to land. To return to our Solar system, they must go through the same process again.

Suppose it takes a year to accelerate and a year to decelerate on the outward journey, and the same on the return journey. If we ignore the time-lapse during coasting as very small in comparison, we might say that it would take a spaceship a total of four years to make the round trip, plus whatever time the astronauts spend exploring their objective. This would be true however distant their destination.

There is a catch, though: the time-rate would slow down only for the astronauts. In the Solar system, where all velocities of material objects are tiny fractions of the speed of light, time proceeds at its customary rate. This means that astronauts returning from Alpha Centauri feeling that four years have passed may find that ten years have passed on Earth. Returning from Vega, again in four years, they may find that perhaps sixty years have passed on Earth. If they go to Rigel and return in the same four years, they may find that over a thousand years have passed on Earth.

(This, by the way, is also true of any scheme to freeze astronauts. To the astronauts, little time-lapse will be experienced, but on Earth the time-lapse will be fully experienced.)

The psychological difficulty for the astronauts, spending years on a spaceship and returning to find that their own world has forever vanished, would be enormous. Near-light-travel may make round-trip visits to the very nearest stars practical, but that practicality rapidly diminishes with distance.

Presumably, then, for serious exploration of anything but our own immediate neighborhood of the Galaxy, for anything but the fifty nearest stars of the hundred billion and more that exist in our Galaxy alone, we must abandon the notion of round trip. Still, the exploration of our own immediate neighborhood is considerably better than nothing. Useful information will surely be gathered and there is a chance that interesting planetary systems may be pinpointed.

Then suppose that colonies are established on some planet of a

Metropolis of the future.

nearby star. These can then eventually serve as a base for explorations still farther out. Then a still-newer colony will serve as a still-newer base. Mankind will creep outward, star-system by star-system. No one group will be able to study, firsthand, more than its own immediate neighborhood among the stars, but mankind as a whole may spread outward without limit.

Which brings us to the matter of the colonization of the planetary systems of other stars.

A colony on the planet of another star cannot count on a lifeline between itself and home. The travelers must carry all that is necessary to make a go of a colony from the start. There must be a wide variety of human genes and human skills (hence perhaps thousands of people to begin with), together with supplies and machinery of all kinds. If we assume that the crew is made up of Earth-men moving outward at last to colonize a planet like their own, we might imagine they would also take selected plants and animals with which to start a self-sufficient ecology.

What we will have is a space ark, so to speak—a huge ship that will itself be a self-contained little universe, capable of supporting itself indefinitely in space.

Can a huge space ark be accelerated to near-light velocities in order to make the space journey seem short to the passengers? Yes, if one is willing to expend the energy required. The more energy used up in acceleration, the less there will be to maintain the ship, and it is very likely that in order to provide for the enormous payload a space ark ought to carry, it will prove impractical to engage in long-drawn-out accelerations and decelerations. Probably the limited energy will be spent in only a short period of acceleration (and later, deceleration); the ark will attain a low velocity and then coast, perhaps for generations.

The difficulties in the way of such an enormous space ark to colonize other star-systems are tremendous. Let us consider them backwards in time. Suppose the Earth-like planet of another star system is reached. Such a planet is very likely to bear life and an independent ecology (though perhaps the chance of *intelligent* life on any given planet is small and need not be taken into account).

Even non-intelligent life raises problems. Is any of it dangerous, either openly or subtly? Will colonists face the possibility of strange diseases, for instance?

We might argue that by the time mankind has developed the technology that makes interstellar travel feasible, it has nothing to fear from any non-intelligent form of life. If that is so, how should the alien ecology best be put to human use? How long ought it to be studied? If the other-planet life-forms are inedible, is it ethical to destroy them and replace the planet's ecology with one of our own? If we put ethics to one side on the basis that necessity knows no law, can an Earth-like ecology be established?

It may well be more difficult to establish a colony against competing life-forms, even if unintelligent, than on a barren world.

Even if we shrug off the difficulties of actually establishing the colony, we must consider the difficulty of spending generations of time in reaching the destination. Will Earth-men be willing to enter a spaceship, knowing that they, and their children and grandchildren, perhaps, will not see the destination but must live out their lives on the artificial world of the ship? Even if some volunteer, will they really be able to bear up through the long years when it is impossible to change their minds and turn back?

Moving backward again, will the people of Earth be willing to invest in huge space arks, requiring enormous investments in effort and resources, in order to send the people aboard them into space to disappear forever—never be heard of again even, very likely, if the trip is successful.

Somehow, all the problems together—building the liners, manning them, sending them through space for generations, colonizing the worlds at the other end—seem to pile up to the point where one cannot be at all confident that this kind of colonization will come to pass.

But so far I have been discussing the colonization of other star systems in terms of high-gravity Earth-men searching for Earth-like planets to colonize. Consider, instead, the low-gravity men of the Moon and other such worlds, and the zero-gravity men of the asteroids.

Here some of the problems disappear, or at least become far less important. The low-gravity and zero-gravity men, used to living inside engineered worlds, will not find space arks a hostile environment and might well have no trouble enduring generations of travel. A man who finds it normal to live out his life in the interior of Ceres can scarcely find it abnormal to live it out in the interior of a space ark.

Then, once the destination is reached, there is a *much* greater likelihood of finding small worlds that will offer acceptable environments within an engineered interior than of finding an Earth-like planet with an acceptable surface. The requirements for Moon-likeness or asteroid-likeness are far less rigid than for Earth-likeness.

Furthermore, there is very little chance of a flourishing ecology on a small body of Moon-size or less. The low-gravity men can take over without problems of either a practical or ethical nature. (And if the planetary system boasts an Earth-like planet, that can be explored to the great advantage of scientific knowledge, without the explorers feeling the urge to occupy the planet, disrupting its life-blanket.)

But there remains the problem of how the space arks could be built on the low-gravity worlds. These are, by definition, worlds with no resources to spare. How could their people find the resources to build huge space arks? Could they turn to Earth? But if Earth were unwilling to make the investment for its own people (as it very well may be), is it not still more unlikely that Earth would be willing to do so for the low-gravity cousins?

Secondly, the volatiles of the comet-asteroid will supply the home-asteroid with enough of the important life elements to replace the leaks in recycling for an indefinite period.

Then, too, the comet-asteroid may offer exercise, adventure, delight. It will be essential to mine the comet-asteroid, and that will mean leaving the "world" and engaging in space activities. Not only will that be a welcome change, but it will keep the population fully aware of the existence of the outside Universe and prevent them from falling into the habit of thinking of their asteroid as all there is.

Perhaps no generation will pass without some small bodies looming far off on the space horizon. Perhaps some of the bodies will not be so small, but may be Moon-sized or even planet-sized bodies circling enormously distant stars, or circling the center of the Galaxy independently. The asteroid-ark, having no specific destination, and having no deadline, will very likely adjust course in order, eventually (and it may take years to reach a body after it has first been detected) to have a chance to examine anything that is encountered.

If the body is small enough to have a negligible gravity, it could be mined for anything that might prove useful, or even be taken along to serve as a back-up or a substitute for comet-asteroids that were being played out. If the body is large enough for its gravity to be uncomfortably strong, the asteroid-ark could still satisfy its curiosity by investigating from afar.

When an ark reaches the neighborhood of a star, with its lighted planets, observations might be particularly intense and particularly interesting. After all, nothing like it may have been seen for thousands of years, and the tale that the asteroid's original home had been in such a system might by then have been clouded into legend, only half-believed.

The asteroid-men might approach close enough to the large planets to study the surface in a rather rapid pass through the upper atmosphere.

How interesting if there were evidence of intelligent life on the planet. There might then be the urge to remain for quite a while to try to piece out the technological level of its inhabitants—what they are, what they do. Naturally, the zero-gravity travelers would never land on these high-gravity planets, and they might be reluctant to complicate matters by establishing communication with the native life-forms. Eventually, they would go off—while the native races might talk excitedly of flying saucers.

And what if the asteroid encountered a star with an asteroid belt— perhaps a not-very-uncommon phenomenon? In that case, a landfall might, in a sense, be made. The asteroid-ark could take up some appropriate orbit and remain there, for some centuries perhaps. Parties would gradually colonize and engineer the other asteroids.

Eventually, after centuries, one or more of the asteroids—or all of them—would set off as asteroid-arks themselves. Perhaps the old, old original ark, worn past repair, would be abandoned—undoubtedly with much more trauma than ever the Sun and Earth were abandoned.

In fact, there might be an "alternation of generations" over the eons as far as the asteroid-arks were concerned. There would be a motile generation in which the arks would move steadily across the vastness of space, but in which population increase in each ark would have to be tightly controlled. Then, after an asteroid belt was encountered, there would be a sessile generation, when for a long period of time there would be no motion but the population would proliferate.

With the conclusion of each sessile generation, there would be a multiplication of asteroid-arks. As the years passed and lengthened into the millions, the arks would begin to swarm over the Universe—all of it their home, all memory of an original Earth vanishing into the mists of forgotten myth.

And what would happen if two asteroid-arks came within detection distance? That, I imagine, would involve a ritual of incomparable importance. There would be no flash-by with a hail and farewell. The arks, having contacted each other as part of a continuing deliberate search over vast distances, would, eventually, be brought to a stable relation to each other and preparations would be made for a long stay.

Each would have compiled its own records, which it could now make available to the other. There would be descriptions, by each, of sectors of space never visited by the other. New theories and novel interpretations of old ones would be expounded. Literature and works of art could be exchanged, differences in custom explained.

167

Most of all there would be the opportunity for a cross-flow of genes. An exchange of population (either temporary or permanent) might be the major accomplishment of any such meeting.

Yet it may happen that such cross-flows will become increasingly impossible. Long isolation may allow the development of human varieties that will no longer be interfertile. The meeting of arks will have to endure long enough, certainly, to learn whether the two populations are compatible. If they are not, intellectual cross-fertilization, at any rate, will be carried on.

But can we consider ourselves the only intelligent race in the Universe capable of launching itself into interstellar space? What if our asteroid belt is already occupied by others? What if we find we are penned into our own inner Solar system with everything beyond blocked off? What if, even before we can get out to the asteroid belt, we begin to notice asteroids moving away as another sessile generation comes to an end? After they are gone, our asteroid belt may be depleted into uselessness.

Very unlikely, of course, on the basis of chance alone. Space arks, moving randomly, are not very likely to stumble upon our star, and the likelihood is that we have escaped. Even if some asteroid-ark of another system has encountered our planetary system, any race capable of interstellar travel may have enough respect for intelligent life to avoid exploiting the asteroid belt of any star that possesses a planet with intelligent life. Perhaps no more than one or two particularly desirable asteroids would be appropriated to replace their worn-out home.

Presumably mankind's own asteroid-arks will practice similar restraint.

But then what if a human asteroid-ark encounters another asteroid-ark and finds, on a close approach, that it is occupied by intelligent beings of utterly different antecedents? Non-human altogether. What then?

All the better perhaps. A genetic interchange would be out of the question, of course, but might not the intellectual cross-fertilization be all the richer, if it could be managed at all?

When humans venture into the wide Universe they may find it already inhabited by a vast brotherhood that does not take into account the differences in the material body, but asks only that the organization of that body be complex enough to support a high intelligence—as it must in that brotherhood, or the asteroid-ark would not have been possible in the first place.

And as part of this vast brotherhood of intelligence, spanning and filling the Universe, mankind might find its true goal at last.

6 Speculations on Another Reality

preceding page: City in the sky. Since anything man can imagine begins to seem possible, perhaps the time will come when he can control the effects of gravity. Then this incredible metropolis suspended among the clouds could become a reality. The city would be largely self-sufficient: here there is a transportation center in the tower and a park area under the transparent dome; apartment complexes along the lower level might house a population of half a million.

opposite: Transportation center floating over the Arizona mountains. Like the preceding, a vision of the distant future, when a great city or a specialized facility like this might be suspended in space wherever needed, oriented in the most effective direction, and moved to another area if required. This transportation port would maintain spacecraft and aircraft, and like a conventional transportation center of today would include services—hotels and restaurants—for passengers. Such floating communities would be largely independent of the Earth's ecology, and the scarred surface of our planet might one day begin to repair itself. (Collection First National Bank of Arizona)

opposite, above: This space robot, its weird appendages designed to perform many tasks, is operated by one man seated safely behind an instrument console at some distant base. Through television and exotic telemetry he can manipulate the robot with great precision.

opposite, below: This fantastic-looking vehicle, a huge steel sphere assembled in space, would propel itself by small atomic charges. A jet of water injected into the chamber forms a hot gas that helps propel the ship, leaving a trail of puffs like those behind a similar craft passing the Moon at the bottom of the painting. The crew rides in the triangular segment in front, which detaches and returns to Earth, leaving the mother ship in orbit to wait for a new nose and a fresh crew.

this page: The space sail—a craft of the distant future for flights to the outer solar system. The space ship dreamed up by H.G. Wells for *First Men on the Moon*, made of a sheet of metal impervious to gravity, might have looked like this. But the space sail would operate on a more rational principle: it would be powered by solar radiation impinging on its vast surface (1500 feet, tip to tip), accelerating slowly, theoretically attaining ultimately the speed of light. The ship would be directed on course by tacking to the Sun, as a sailboat tacks to the wind.

right: The plausible can carry us just so far—at some point we must enter the far-out dream in which we might find ourselves if a time machine could transport us two centuries hence. Here is a suggestion of the strange world man might encounter if he visited another planetary system of our galaxy. There is no sense of up or down, no gravity effect. A beam of diffused light is perhaps transporting ionized particles of material objects across vast distances.

next page: The surface of the Moon, the globe of the Earth, the Sun. Copernicus taught us that we are not the center, but until men stood on the Moon and looked back to the small sphere of Earth we did not quite believe it. By 2200 we may have explored and colonized our solar system and have learned more of man's capacities and his place in the universe.

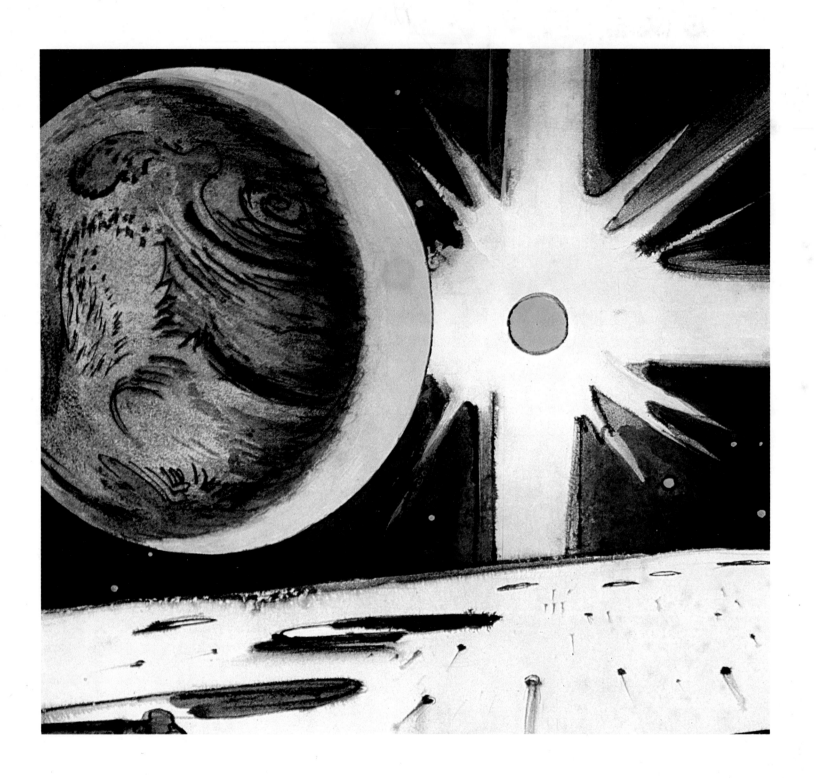